ISSN 0044-2267

ZEITSCHRIFT FÜR ANGEWANDTE MATHEMATIK UND MECHANIK

INGENIEURWISSENSCHAFTLICHE FORSCHUNGSARBEITEN

BAND 61 1981 HEFT 12

AUS DEM INHALT

HAUPTAUFSÄTZE

KLEINE MITTEILUNGEN

BUCHBESPRECHUNGEN

THEORIA CUM PRAXI

AKADEMIE-VERLAG · BERLIN

ZAMM Bd. 61 Nr. 12 S. 605—668 Berlin, Dezember 1981

EVP 18,— M 34115

INHALT

Wir bitten, **Manuskriptsendungen** zweifach (Original und eine Kopie, sprachlich einwandfrei, Formeln mit Maschine oder in Druckschrift geschrieben) an folgende Anschrift zu richten:

Zeitschrift für Angewandte Mathematik und Mechanik,
Institut für Mechanik der Akademie der Wissenschaften der DDR
DDR-1199 Berlin, Rudower Chaussee 5.

Zu den Arbeiten, die als Hauptaufsätze bestimmt sind, ist auf gesondertem Blatt eine Zusammenfassung von 5 bis 10 Zeilen in englischer und (möglichst) deutscher und russischer Sprache beizufügen. Ausführliche Hinweise für die Autoren, um deren strikte Berücksichtigung gebeten wird, finden sich im Anschluß an das Inhaltsverzeichnis des Jahrganges **60** (1980).

Die Autoren erhalten von den Hauptaufsätzen 75, von den Kleinen Mitteilungen 25 Sonderdrucke ohne Berechnung, darüber hinaus weitere Sonderdrucke gegen Berechnung.

Der Verlag behält sich für alle Beiträge das Recht der Vervielfältigung und Übersetzung vor.

Bestellungen sind zu richten

- **in der DDR** an den Postzeitungsvertrieb unter Angabe der Kundennummer des Bestellenden oder an den AKADEMIE-VERLAG, DDR-1086 Berlin, Leipziger Straße 3—4
- **im sozialistischen Ausland** an eine Buchhandlung für fremdsprachige Literatur oder an den zuständigen Postzeitungsvertrieb
- **in der BRD und Berlin(West)** an eine Buchhandlung oder an die Auslieferungsstelle KUNST UND WISSEN, Erich Bieber OHG, D-7000 Stuttgart 1, Wilhelmstraße 4—6
- **in den übrigen westeuropäischen Ländern** an eine Buchhandlung oder an die Auslieferungsstelle KUNST UND WISSEN, Erich Bieber GmbH, Ch-8008 Zürich/Schweiz, Dufourstraße 51
- **im übrigen Ausland** an den Internationalen Buch- und Zeitschriftenhandel; den Buchexport, Volkseigener Außenhandelsbetrieb der Deutschen Demokratischen Republik, DDR-7010 Leipzig, Postfach 160; oder an den AKADEMIE-VERLAG, DDR-1086 Berlin, Leipziger Straße 3—4

ZEITSCHRIFT FÜR ANGEWANDTE MATHEMATIK UND MECHANIK

Herausgeber und Chefredakteur: Prof. Dr. Günter Schmidt. Redaktion: Dr. Winfried Heinrich, Dr. Horst Weinert, Dipl.-Math. Friedhild Dudel, Helga Rühl, Institut für Mechanik der Akademie der Wissenschaften der DDR. Verlag: Akademie-Verlag, DDR-1086 Berlin, Leipziger Straße 3—4; Fernruf: 2236221 oder 2236221. Telex-Nr.: 114420; Bank: Staatsbank der DDR, Berlin, Kto.-Nr.: 6836-26-20712. Anschrift der Redaktion: Institut für Mechanik der Akademie der Wissenschaften, DDR-1199 Berlin, Rudower Chaussee 5, Fernruf: 6702841.

Veröffentlicht unter der Lizenznummer 1282 des Presseamtes beim Vorsitzenden des Ministerrates der Deutschen Demokratischen Republik. Gesamtherstellung: VEB Druckerei „Thomas Müntzer", DDR-5820 Bad Langensalza. Erscheinungsweise: Die Zeitschrift für Angewandte Mathematik und Mechanik erscheint monatlich. Die 12 Hefte eines Jahres einschließlich Tagungshefte bilden einen Band. Bezugspreis je Band 360,— M zuzüglich Versandspesen (Preis für die DDR 216,— M). Bezugspreis je Heft 30,— M (Preis für die DDR 18,— M). Bestellnummer dieses Heftes: 1009/61/12.

AN (EDV) 35937

ZAMM 61, 605 – 615 (1981)

P. Lötstedt

Coulomb Friction in Two-Dimensional Rigid Body Systems

Die Lösungen des Coulombschen Reibungsproblems für starre Körper in zwei Dimensionen werden analysiert. Das bestimmende System von gewöhnlichen Differentialgleichungen und Ungleichungen wird aufgestellt. Beispiele werden vorgelegt, die einige ungewünschte Eigenschaften dieses speziellen Reibungsgesetzes nachweisen. Hinreichende Bedingungen für Existenz und Eindeutigkeit werden mit Hilfe der Theorie der linearen Komplementarität hergeleitet.

The solutions to the Coulomb friction problem for rigid bodies in two dimensions are analyzed. The governing system of ordinary differential equations and inequalities is derived. Examples are presented demonstrating undesirable properties of this particular law of friction. Sufficient conditions for existence and uniqueness are given using the theory of linear complementarity.

Анализируются решения двумерной проблемы трения Кулона твёрдых тел. Строится соответствующая система обыкновенных дифференциальных уравнений и неравенств. Приводятся примеры, демонстрирующие некоторые нежелательные свойства этого закона трения. Даются достаточные условия существования и единственности решения на основе теории линейной дополнительности.

1. Introduction

In this paper we are concerned with the existence and uniqueness of solutions of the two-dimensional Coulomb friction problem for systems of rigid bodies. It was first pointed out by Painlevé [18] that there are simple examples where no solution exists and if it exists it is not necessarily unique. This fact has since been discussed by a number of authors, e.g. Béghin [1], [2], Delassus [7], Klein [13] and Pérès [19]. In particular in the numerical simulation of the motion of rigid body systems subject to Coulomb friction (see Cundall and Stark [6] and Lötstedt and Dahlquist [15] for examples), it is of importance to be aware of the deficiencies of the mathematical model. Therefore, sufficient conditions for existence and uniqueness of a solution will be derived here using the theory of linear complementarity from mathematical programming, Cottle and Dantzig [5] and Karamardian [11].

Let two rough, rigid bodies in a planar system be in contact with each other at a point. Denote the relative velocity between the bodies at the point of contact by v, the normal force by N and the friction force by F (cf. fig. 1). Then, according to Coulomb's law, Kilmister and Reeve [12] or Pérès [19],

$$1.\quad N \geqq 0\,, \qquad 2.\quad \text{if } v \neq 0 \text{ then } |F| = \mu N \text{ and } vF \leqq 0\,, \qquad 3.\quad \text{if } v = 0 \text{ then } |F| \leqq \mu N\,, \tag{1.1}$$

where $\mu > 0$ is the friction coefficient. In order to state the equations and relations to be satisfied by a general system, let $q \in R^m$ be the coordinate vector of the system and let

$$\phi_i(q) \geqq 0\,, \qquad i = 1, 2, \dots, p\,, \tag{1.2}$$

be the unilateral constraints associated with the normal forces. A constraint function ϕ_j may e.g. be the expression for the distance between the corner of one body and the edge of another body. Furthermore, let $g_{Fi}(q) \in R^m$, let the superscript T denote transposition and let

$$g_{Fi}^T\dot{q},\ i = 1, 2, \dots, k \leqq p\,, \tag{1.3}$$

be the relative velocities between the bodies at the points of contact in the tangential direction. Assume that the ϕ_j and $g_{Fj}^T\dot{q}$ are numbered so that those with the same index j belong to the same point of contact. Smooth contacts without friction have indices $k+1, k+2, \dots, p$. The following sets will simplify the notation in the sequel,

$$\begin{aligned} J_N &= \{i \mid \phi_i = 0\}\,, \\ J_{F0} &= \{i \mid i \in J_N,\, g_{Fi}^T\dot{q} = 0\}\,, \qquad J_{FW} = \{i \mid i \in J_N,\, g_{Fi}^T\dot{q} \neq 0\}\,, \\ J_F &= J_{F0} \cup J_{FW}\,. \end{aligned} \tag{1.4}$$

Introduce

$$g_{Ni}(q) = \frac{\partial \phi_i}{\partial q} \in R^m \tag{1.5}$$

and suppose that $J_{FW} = \emptyset$. Then $g_{Ni}^T\dot{q} = 0$, $i \in J_N$, and $g_{Fi}^T\dot{q} = 0$, $i \in J_F$. The equations of motion for a planar system of rigid bodies are, Pérès [19],

$$M(q)\,\ddot{q} = f(q, \dot{q}, t) + \sum_{i \in J_N} g_{Ni}\lambda_{Ni} + \sum_{i \in J_{F0}} g_{Fi}\lambda_{Fi}\,, \tag{1.6}$$

where $M \in R^{m \times m}$ is symmetric and positive definite almost everywhere. $f \in R^m$ contains the external forces and inertia forces. λ_{Ni} and λ_{Fi} are the Lagrange multipliers associated with the constraints $\phi_i = 0$ and $g_{Fi}^T\dot{q} = 0$. $g_{Ni}\lambda_{Ni}$ can be interpreted as a force acting in R^m. The same interpretation holds for $g_{Fi}\lambda_{Fi}$. With an appropriate scaling of g_{Ni} and g_{Fi}, λ_{Ni} is proportional to N and λ_{Fi} to F with the same proportionality factor.

Let the coordinates of a body be the Cartesian coordinates of the center of gravity and an angle for the orientation of the body. Then M is constant and diagonal with the positive masses and moments of inertia in the diagonal elements. Without loss of generality, we can assume that $M = I$, the identity matrix, since

$$M = C^2, \quad q^* \equiv Cq, \quad \phi_i^*(q^*) \equiv \phi_i(C^{-1}q^*) = \phi_i(q), \quad f^*(q^*, \dot q^*, t) \equiv C^{-1}f(q, \dot q, t),$$
$$g_{Ni}^*(q^*) = \frac{\partial \phi_i^*}{\partial q^*} = C^{-1}g_{Ni}(q), \quad g_{Fi}^*(q^*) \equiv C^{-1}g_{Fi}(q).$$

Premultiply (1.6) by C^{-1} and we obtain the equation for q^* which is equivalent to (1.6) with $M = I$.

We shall continue with a discussion what happens when an index j is dropped from J_N, i.e. ϕ_j becomes positive, and when an index is transferred from J_{F0} to J_{FW}. Suppose that in $[0, t_0]$, $t_0 > 0$, $\phi_j = 0$ but for $t > t_0$, $\phi_j > 0$. If the first and second derivatives of ϕ_j are continuous in $[0, t_0]$, then

$$\frac{d}{dt}\phi_j = g_{Nj}^T\dot q = 0, \quad \delta_j = \frac{d^2}{dt^2}\phi_j = g_{Nj}^T\ddot q + \dot g_{Nj}^T\dot q = 0,$$

but at least for a short interval $(t_0, t_0 + \Delta t')$, $\Delta t' > 0$, $g_{Nj}^T\dot q > 0$ and for $(t_0, t_0 + \Delta t)$, $0 < \Delta t < \Delta t'$, $\delta_j > 0$. λ_{Nj} is positive (or at least non-negative) as long as $\phi_j = 0$ but $\lambda_{Nj}(t_0) = 0$ which prevails until $\phi_j = 0$ again. By expanding ϕ_j in a Taylor series we conclude that λ_{Nj}, ϕ_j, $g_{Nj}^T\dot q$ and δ_j satisfy the complementarity conditions:

$$\begin{aligned}&\lambda_{Nj} \geqq 0, \quad \phi_j \geqq 0, \quad \lambda_{Nj}\phi_j = 0, \quad t \in [0, \infty),\\ &g_{Nj}^T\dot q \geqq 0, \quad \lambda_{Nj}g_{Nj}^T\dot q = (g_{Nj}\lambda_{Nj})^T\dot q = 0, \quad t \in [0, t_0 + \Delta t'],\\ &\delta_{Nj} \geqq 0, \quad \lambda_{Nj}\delta_{Nj} = 0, \quad t \in [0, t_0 + \Delta t].\end{aligned} \tag{1.7}$$

Suppose that $j \in J_N$ and $g_{Fj}^T\dot q = 0$ in $[0, t_0]$ but for $t > t_0$ $g_{Fj}^T\dot q \neq 0$. In a manner similar to that above, utilizing the equivalence between $g_{Fj}^T\dot q$ and v in (1.1), the definitions (1.1) and (1.3) and the scaling of g_{Nj} and g_{Fj} the following relations are derived for λ_{Fj}, $g_{Fj}^T\dot q$ and $\delta_{Fj} = g_{Fj}^T\ddot q + \dot g_{Fj}^T\dot q$:

$$\begin{aligned}&-\mu\lambda_{Nj} \leqq \lambda_{Fj} \leqq \mu\lambda_{Nj},\\ &\lambda_{Fj}g_{Fj}^T\dot q = (g_{Fj}\lambda_{Fj})^T\dot q \leqq 0, \quad t \in [0, \infty),\\ &|\lambda_{Fj}| = \mu\lambda_{Nj}, \quad t \in [t_0, t_0 + \Delta t'],\\ &(\mu\lambda_{Nj} - |\lambda_{Fj}|)\,\delta_{Fj} = 0, \quad \lambda_{Fj}\delta_{Fj} \leqq 0, \quad t \in [0, t_0 + \Delta t],\end{aligned} \tag{1.8}$$

for some $\Delta t'$ and Δt, $0 < \Delta t < \Delta t'$.

From the interpretation of $g_{Fj}\lambda_{Fj}$ as a force and (1.8) the conclusion is that the contribution to the right hand side of (1.6) from a working friction force, $j \in J_{FW}$, is

$$g_{Fj}\lambda_{Fj} = g_{Fj}s_j\mu\lambda_{Nj} \tag{1.9}$$

where $s_j = -\text{sign}\,(g_{Fj}^T\dot q)$. Let

$$\begin{aligned}&G = (g_{Ni_1}, \dots, g_{Ni_n}, g_{Fj_1}, \dots, g_{Fj_r}), \quad i_l \in J_N, \quad n \leqq p, \quad j_l \in J_F, \quad r \leqq k,\\ &G_1 = (g_{Ni_1}, \dots, g_{Ni_n}, g_{Fj_1}, \dots, g_{Fj_s}), \quad i_l \in J_N, \quad j_l \in J_{F0}, \quad s \leqq r,\\ &H = (g_{Fj_1}, \dots, g_{Fj_u}), \quad u = r - s, \quad j_l \in J_{FW},\end{aligned} \tag{1.10}$$

and let $\lambda \in R^{n+r}$ and $\lambda_1 \in R^{n+s}$ consist of multipliers corresponding to the columns in G and G_1, respectively. $G\lambda$ is the sum of all constraint forces and $G_1\lambda_1$ the sum of the non-working constraint forces. Define $U \in R^{u\times(n+s)}$ such that (cf. (1.9))

$$\sum_{i \in J_{FW}} g_{Fi}\lambda_{Fi} = HU\lambda_1. \tag{1.11}$$

Suppose that J_N and J_{F0} are constant in $(0, t_0]$ and that no constraint is added to J_N or J_{F0} in $(t_0, t_0 + \Delta t'']$ for $\Delta t'' > 0$ sufficiently small. Then the formulas (1.6), (1.7), (1.8), (1.10), (1.11) (H consists of columns g_{Fi} such that $i \in J_{FW}$ in $(0, t_0]$) and the previous discussion are summarized in a system of ordinary differential equations and relations between the Lagrange multipliers and the relative distances, velocities and accelerations at the points of contact:

$$\ddot q = f + G\lambda \tag{1.12a}$$

or

$$\ddot q = f + (G_1 + HU)\lambda_1, \tag{1.12b}$$

$$\phi_i \geqq 0, \quad \lambda_{Ni} \geqq 0, \quad \lambda_{Ni}\phi_i = 0, \quad \forall i, \quad t \in [0, \infty).$$

If $i \in J_N$ in $(0, t_0]$ then

$$g_{Ni}^T\dot q \geqq 0, \quad \lambda_{Ni}g_{Ni}^T\dot q = 0, \quad t \in (0, t_0 + \Delta t'], \quad \Delta t' < \Delta t''.$$

If $g_{Ni}^T\dot q = 0$ in $(0, t_0]$ then

$$\begin{aligned}&\delta_{Ni} \geqq 0, \quad \lambda_{Ni}\delta_{Ni} = 0, \quad t \in (0, t_0 + \Delta t], \quad \Delta t < \Delta t',\\ &|\lambda_{Fi}| \leqq \mu\lambda_{Ni}, \quad i \in J_F.\end{aligned} \tag{1.12c}$$

If $i \in J_{FW}$ in $(0, t_0]$ then

$$|\lambda_{Fi}| = \mu\lambda_{Ni}, \quad \lambda_{Fi}g_{Fi}^T\dot q \leqq 0, \quad t \in (0, t_0 + \Delta t'']$$

else

$$i \in J_{F0} \text{ in } (0, t_0],$$
$$(\mu\lambda_{Ni} - |\lambda_{Fi}|)\,\delta_{Fi} = 0\,, \quad \lambda_{Fi}\delta_{Fi} \leqq 0\,, \quad t \in (0, t_0 + \Delta t]\,.$$

The reason why the time-interval is not closed at 0 is that we want to allow a constraint to be added to or dropped from J_N or J_{F0} at $t = 0$, e.g. $\phi_i(0) = 0$ but $\phi_i(t) > 0$, $t > 0$.

In (1.12) the governing equations and relations are given in an interval where indices are dropped from J_A and J_{F0}. Define

$$J_N' = \{i \mid \phi_i = 0 \text{ and } g_{Ni}^T\dot{q} = 0\}\,,$$
$$J_{F0}' = J_{F0} \cap J_N'\,, \qquad J_{FW}' = J_{FW} \cap J_N'\,. \tag{1.13}$$

Let J_N', J_{F0}' and J_{FW}' determine G_1 and H as in (1.10). Observe that if $g_{Nj}^T\dot{q}(t_0) > 0$ then $\phi_j(t) > 0$ in (t_0, t_1) for some $t_1 > t_0$ and by (1.7) $\lambda_{Nj}(t_0) = 0$. Let $\delta_1 \in R^{n+s}$ contain those δ_{Ni} and δ_{Fi} which correspond to λ_{Ni} and λ_{Fi} in λ_1 and insert (1.12b) into the definition of δ_1. Then the problem to be solved for λ_1 and λ at a point t_0 is

$$G_1^T(G_1 + HU)\,\lambda_1 + \dot{G}_1^T\dot{q} + G_1^Tf = \delta_1 \tag{1.14a}$$

$$\lambda_{Ni} \geqq 0\,, \quad \delta_{Ni} \geqq 0\,, \quad \lambda_{Ni}\delta_{Ni} = 0\,, \quad i \in J_N' \text{ at } t_0\,,$$
$$-\mu\lambda_{Ni} \leqq \lambda_{Fi} \leqq \mu\lambda_{Ni}\,, \quad i \in J_{F0}' \text{ at } t_0\,,$$
$$\text{if } \delta_{Fi} \neq 0 \text{ then } |\lambda_{Fi}| = \mu\lambda_{Ni} \text{ and } \lambda_{Fi}\delta_{Fi} \leqq 0\,, \tag{1.14b}$$

which is equivalent to the formulation in (1.12). To simplify the notation it is henceforth assumed that $J_N' = \{1, 2, \ldots, n\}$.

The sections 2, 3 and 4 are devoted to an investigation of the existence and uniqueness of the solution of (1.14) at a point t_0. The properties of the time-dependent solution (1.12) and what happens when constraints are added to J_N and J_{F0} are discussed in the last section.

2. Examples of non-existence and non-uniqueness

Three examples will be studied, which exhibit undesirable behaviour of Coulomb's law of friction.

Example 1: Existence

Let us consider the rigid rod and the floor in fig. 1. The example resembles one investigated by BÉGHIN [1] and KLEIN [13]. In this particular case

$$g_N = \begin{pmatrix} 0 \\ 1 \\ -l\cos\varphi \end{pmatrix}, \quad g_F = \begin{pmatrix} 1 \\ 0 \\ l\sin\varphi \end{pmatrix}, \quad q = \begin{pmatrix} x \\ y \\ \varphi \end{pmatrix}. \tag{2.1}$$

Suppose that at $t = t_0$, e.g. after a collision with the floor $g_N^T\dot{q} = 0$ and $g_F^T\dot{q} < 0$. By (1.12c) we have $\lambda_F = \mu\lambda_N$ and the equation corresponding to (1.12b) is

$$\ddot{q} = \begin{pmatrix} f_x \\ f_y \\ f_\varphi \end{pmatrix} + \begin{pmatrix} \mu \\ 1 \\ -l\,(\cos\varphi - \mu\sin\varphi) \end{pmatrix}\lambda_N\,. \tag{2.2}$$

In accordance with (1.12) we have taken the mass m and the moment of inertia J of the rod such that $m = J = 1$. With mass symmetrically distributed about the geometrical midpoint of the rod, l cannot be chosen arbitrarily.

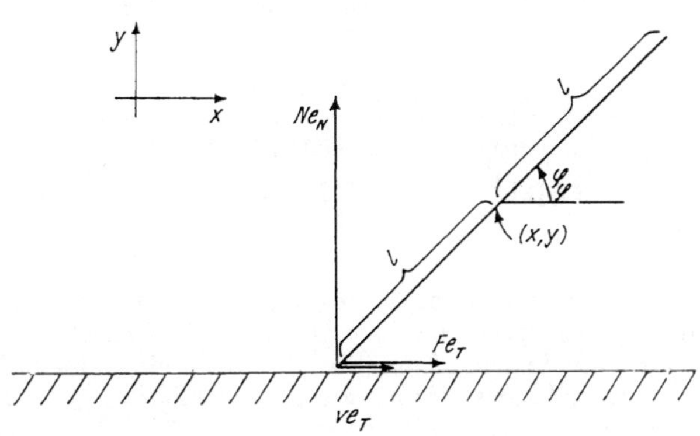

Fig. 1. A thin, rigid rod on a rough floor. e_T and e_N are unit vectors.

It is easily shown that $l \geqq 1$. The equation to be satisfied by λ_N at t_0 is (cf. (1.14))

$$a\lambda_N + b = \delta_N \geqq 0, \quad \lambda_N \geqq 0, \quad \lambda_N\delta_N = 0, \quad a = 1 + l^2 \cos\varphi(\cos\varphi - \mu\sin\varphi), \tag{2.3}$$
$$b = f_y - f_\varphi l\cos\varphi + l\dot\varphi^2 \sin\varphi,$$

where f_y and f_φ are evaluated at t_0. If $b \geqq 0$ then $\lambda_N = 0$ is a solution and if $b < 0$ and $a > 0$ then $\lambda_N = -b/a$, $\delta_N = 0$ fulfils (2.3). However, when $a < 0$ and $b < 0$ there is no solution λ_N, δ_N satisfying (2.3). A negative b is obtained by taking $\dot\varphi = f_\varphi = 0$ and $f_y = -g < 0$, the gravity force. Are there l, φ and μ such that $a < 0$? Let $\varphi \in (0, \pi/2)$. After elementary calculations one finds that $a(\varphi)$ has one and only one minimum $\varphi_{\min}$ in $(\pi/4, \pi/2)$ and $a(\varphi_{\min}) < 0$ if

$$1 + 2/l^2 < \sqrt{1 + \mu^2}. \tag{2.4}$$

Since $a(\varphi)$ is continuous in φ there is a sector (φ_1, φ_2) in which $a < 0$. (2.4) implies that for every μ there is an l such that $a < 0$ in (φ_1, φ_2). For $\mu = 0.6$, l must satisfy $l \gtrsim 3.47$. It should be noted that for small μ, l and the rod have to be rather extreme.

If $b > 0$ and $a < 0$ then the solution $\lambda_N = 0$ and $\delta_N = b$ is not unique. $\lambda_N = -b/a$ and $\delta_N = 0$ are also possible.

Example 2: Uniqueness

Consider the body in example 1 once more. Suppose that $g_F^T\dot q = 0$. Then $J_{F\dot W} = \emptyset$ and $G_1 = (g_N, g_F) = G_1 + HU$. With this G_1 in (1.14) the definitions of u, v, w, x and y below are obvious. λ_N and λ_F must satisfy

$$u\lambda_N + v\lambda_F + x = \delta_N \geqq 0, \quad v\lambda_N + w\lambda_F + y = \delta_F, \tag{2.5a}$$
$$\lambda_N \geqq 0, \quad \lambda_N\delta_N = 0, \quad |\lambda_F| \leqq \mu\lambda_N; \tag{2.5b}$$
$$\text{if } \delta_F > 0 \text{ then } \lambda_F = -\mu\lambda_N, \quad \text{if } \delta_F < 0 \text{ then } \lambda_F = \mu\lambda_N.$$

Since $G_1^TG_1$ is positive definite, $u > 0$, $w > 0$ and

$$uw - v^2 > 0 \Leftrightarrow u/|v| > |v|/w. \tag{2.6}$$

$v = -l^2 \sin\varphi\cos\varphi$ and since $\varphi \in (0, \pi/2)$, $v < 0$. Choose f, φ, $\dot\varphi$, l and μ such that $x > 0$, $y < 0$ and

$$(-v)/u > 1/\mu \Leftrightarrow u + \mu v < 0. \tag{2.7}$$

Observe that $a = u + \mu v$ in (2.3).

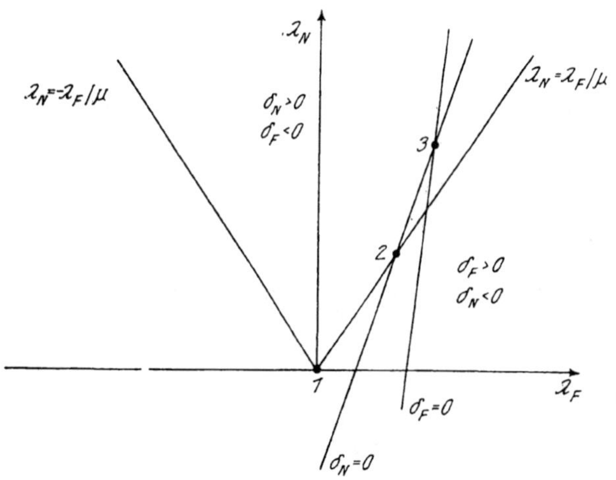

Fig. 2. Illustration of a one-contact system with three possible solutions

Fig. 2 is an illustration of (2.5a) with u, v, w, x and y chosen as above. The solution must be sought in the sector between $\lambda_N = \lambda_F/\mu$ and $\lambda_N = -\lambda_F/\mu$ and on or above the line $\delta_N = 0$ in order to comply with (2.5b). By inspection of fig. 2 we find that three different solutions are possible:

1. $\lambda_N = \lambda_F = 0$, $\delta_N > 0$, the rod leaves the floor,
2. $\lambda_N > 0$, $\lambda_F = \mu\lambda_N$, $\delta_N = 0$, $\delta_F < 0$, the rod starts sliding on the floor,
3. $\lambda_N > 0$, $\delta_N = 0$, $\delta_F = 0$, the rod and the floor remain in contact without sliding.

A further discussion of similar examples is given in BÉGHIN [2] and PÉRÈS [19].

Example 3: Sensitivity to perturbations

An example of the effect of small disturbances is given in fig. 3. It turns out that a small perturbation of x in (2.5a) will cause a large discrepancy between the solution of the perturbed and the unperturbed system. u, v, w and y have the same values as in the previous example. Let the subscripts a and b on a variable denote its value in fig. 3a and fig. 3b, respectively. Here $|x/v|$ is small, $x_a > 0$ and in fig. 3b x is perturbed slightly so that $x_b < 0$. If we follow a suggestion by DELASSUS, see PÉRÈS [19], and always choose $\lambda_N = 0$ if possible, then in fig. 3a at point 1, λ_{Na}

$= \lambda_{Fa} = 0$, $\delta_{Na} > 0$ is the solution. In fig. 3b, $\lambda_{Nb} > 0$, $\lambda_{Fb} > 0$, $\delta_{Nb} = \delta_{Fb} = 0$ is the only solution at point 3. Hence, $|\lambda_{Na} - \lambda_{Nb}|$ does not depend continuously on $|x_a - x_b|$. The choice of λ_N and λ_F at point 2 does not improve the situation. If we, contrary to DELASSUS, always take the solution $\lambda_N \neq 0$ (point 3), the same conclusion can be drawn from another example as simple as that in fig. 3.

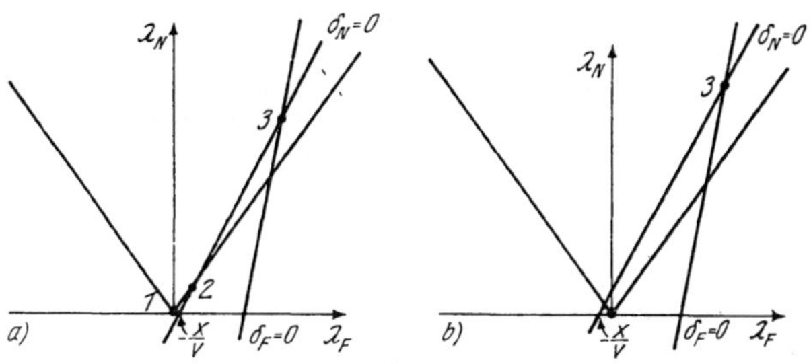

Fig. 3. Illustration of the effect of a small disturbance in a one-contact system

3. Sufficient conditions for existence

In the previous section an example was given for which no allowable pair (λ_N, δ_N) could be determined at a time point t_0. To begin with, two special cases when a solution always exists will be analyzed and then sufficient conditions for existence of a solution of (1.14) are presented.

Assume that every λ_{Ni} is known independent of λ_{Fi}, e.g. as a function of q, $\dot{q}$ and t. The contribution of the multipliers to the equations of motion (1.12a) $G\lambda$, is split into three parts

$$G\lambda = G_N\lambda_N + HU\begin{pmatrix}\lambda_N\\0\end{pmatrix} + G_F\lambda_F \tag{3.1}$$

where

$$G_N = (g_{N1}, \dots, g_{Nn}), \qquad J_N' = \{1, 2, \dots, n\} \text{ at } t_0, \qquad G_F = (g_{Fi_1}, \dots, g_{Fi_s}), \qquad i_j \in J_{F0}' \text{ at } t_0.$$

λ_N and λ_F contain the multipliers associated with the columns in G_N and G_F. Note that

$$U\begin{pmatrix}\lambda_N\\0\end{pmatrix} = U\begin{pmatrix}\lambda_N\\\lambda_F\end{pmatrix}.$$

Thus, (1.12a) can be written

$$\ddot{q} = F(q, \dot{q}, t) + G_F\lambda_F$$

with an obvious definition of F and consequently, (1.14) has the simplified form

$$G_F^TG_F\lambda_F + \dot{G}_F^T\dot{q} + G_F^TF = \delta_F, \qquad \delta_F = (\delta_{Fi_1}, \dots, \delta_{Fi_s}), \qquad i_j \in J_{F0}; \tag{3.2}$$

$$-\mu\lambda_{Ni} \leqq \lambda_{Fi} \leqq \mu\lambda_{Ni}, \qquad \text{if } \delta_{Fi} \neq 0 \text{ then } |\lambda_{Fi}| = \mu\lambda_{Ni} \text{ and } \lambda_{Fi}\delta_{Fi} \leqq 0.$$

The conditions in (3.2) are equivalent to the Kuhn-Tucker conditions, see e.g. POWELL [20], satisfied by the optimal solution of the quadratic programming problem

$$\min_{\lambda_F} \tfrac{1}{2}\lambda_F^TG_F^TG_F\lambda_F + \lambda_F^T(\dot{G}_F^T\dot{q} + G_F^TF); \qquad -\mu\lambda_{Ni} \leqq \lambda_{Fi} \leqq \mu\lambda_{Ni}. \tag{3.3}$$

According to ORTEGA and RHEINBOLDT [17, 4.2.2] a solution λ_F^* to (3.3) always exists. If G_F has full column rank then a convex function is minimized over a convex set and λ_F^* is unique. Coulomb friction problems in connection with linear elasticity are discussed in DUVAUT and LIONS [8]. Questions of existence and uniqueness are settled only under simplifying assumptions such as those leading to (3.2). The more general problems remain open.

The fact that (3.2) expresses the Kuhn-Tucker conditions for (3.3) will now be exploited further. Let us return to the one-contact problem (2.5) and consider the minimum of

$$\psi(\lambda_N, \lambda_F) = \frac{1}{2}\begin{pmatrix}\lambda_N\\\lambda_F\end{pmatrix}^T\begin{pmatrix}u & v\\v & w\end{pmatrix}\begin{pmatrix}\lambda_N\\\lambda_F\end{pmatrix} + \begin{pmatrix}\lambda_N\\\lambda_F\end{pmatrix}^T\begin{pmatrix}x\\y\end{pmatrix} \tag{3.4}$$

over the set D where

$$D = D_1 \cap D_2 \cap D_3, \qquad D_1 = \{\lambda_N, \lambda_F \mid |\lambda_F| \leqq \mu\lambda_N, \lambda_N \geqq 0\}, \tag{3.5}$$

$$D_2 = \{\lambda_N, \lambda_F \mid \text{if } x > 0 \text{ then } \lambda_N = \lambda_F = 0 \text{ else if } -u + \mu v < 0 \text{ then } \lambda_F \geqq -\mu x/(-u + \mu v) \text{ else } -\infty < \lambda_F < \infty\},$$

$$D_3 = \{\lambda_N, \lambda_F \mid \text{if } x > 0 \text{ then } \lambda_N = \lambda_F = 0 \text{ else if } u + \mu v > 0 \text{ then } \lambda_F \leqq -\mu x/(u + \mu v) \text{ else } -\infty < \lambda_F < \infty\}.$$

The set D is depicted in fig. 4 for $u > \mu|v|$ and $-\mu v > u > 0$. The following well-known result, see e.g. POWELL [20], characterizes the minimum of a differentiable function.

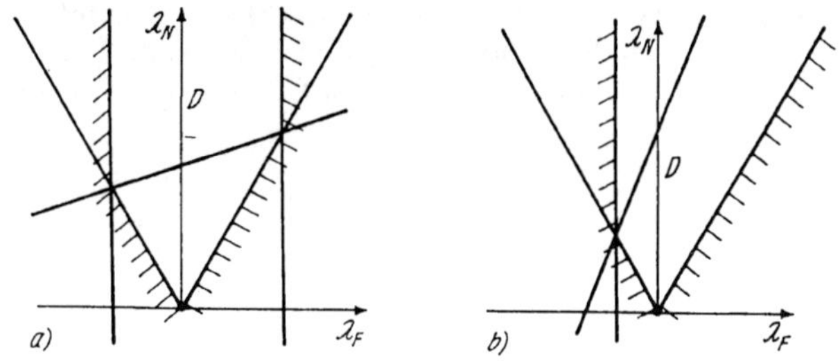

Fig. 4. The region D in (3.5). In 4a $u > \mu |v|$ and in 4b $-\mu v > u > 0$

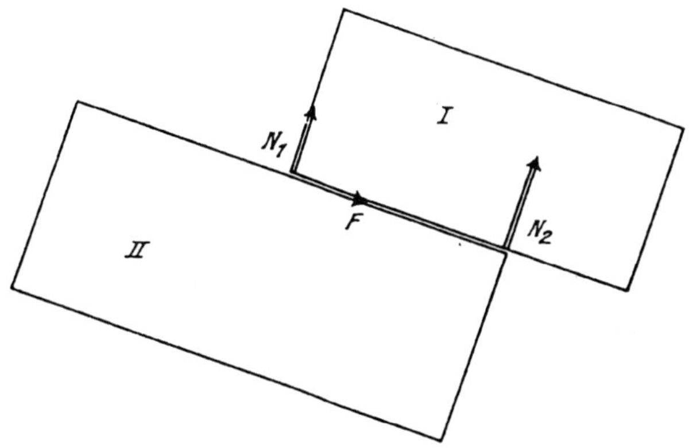

Fig. 5. An edge-to-edge contact between two bodies. N_1 and N_2 represent the normal forces and F the friction force on body I

Lemma 1.

Suppose that $f(z)$ has at least one derivative ∇f. A necessary condition for z^* to solve the constrained minimization problem

$$\min_{x \in S} f(z)$$

is that there is no vector p such that

$$(\nabla f(z^*))^T p < 0$$

and $z^* + \alpha p \in S$ for all values of α in $0 < \alpha < \bar{\alpha}$.

Apply the lemma to the above problem (3.4), (3.5). It is easily verified that the minimum of (3.4) over D satisfies (1.14) for this one-contact system. In general, the minimum of (3.4) over D_1 does not fulfil (1.14). The solution to the problem illustrated in fig. 2 will be $\lambda_N = \lambda_F = 0$ since $D = (0, 0)$.

This approach can be generalized to a specific n-contact system, viz. if $J'_{FW} = \emptyset$ at t_0 and the columns of G are linearly independent. Then there is no matrix H in (1.14a) and inserting $G_1 = G$, $\lambda_1 = \lambda$ and $\delta_1 = \delta$ yields

$$G^T G\lambda + \dot{G}^T \dot{q} + G^T f = \delta \,. \tag{3.6}$$

A purely statical problem can be stated in this form. Suppose that the multipliers are ordered in λ so that

$$\begin{aligned} &\lambda_j = \lambda_{Nj}\,, \quad j = 1, 2, \dots, n\,, \quad \lambda_{n+j} = \lambda_{Fj}\,, \quad j = 1, 2, \dots, s \leqq n\,. \\ &\lambda_{Fj} = \lambda_{n+j} \text{ is associated with } \lambda_{Nj} = \lambda_j\,. \end{aligned} \tag{3.7}$$

Let $A = G^T G$ and introduce $a_i \in R^{n+s}$, $i = 1, 2, \dots, s$, where

$$a_{ij} = \begin{cases} A_{ij}\,, & j \neq i\,, \quad j \neq n + i \\ 0, & j = i\,, \quad j = n + i \end{cases}$$

and

$$b_i = a_i^T \lambda + (\dot{G}^T \dot{q} + G^T f)_i \in R\,.$$

The definitions of the sets D_1^i, D_2^i and D_3^i, $i \leqq s$, are the same as those of D_1, D_2 and D_3 where λ_N, λ_F, x, u and v are replaced by λ_i, λ_{n+i}, b_i, A_{ii} and $A_{i,n+i}$, respectively.

$$D^i = D_1^i \cap D_2^i \cap D_3^i\,, \qquad i = 1, 2, \dots, s\,,$$

and for $s + 1 \leqq i \leqq n$

$$D^i = \{\lambda_{Ni} \mid \lambda_{Ni} \geqq 0\}\,.$$

We shall now show that the optimal point of a minimization problem satisfies (3.6) and (1.14b).

Theorem 1.

Let G have full column rank and define $\Psi(\lambda)$ and D

$$\Psi(\lambda) = \tfrac{1}{2}\lambda^T G^T G\lambda + \lambda^T(\dot{G}^T \dot{q} + G^T f)\,, \qquad D = \bigcap_{i=1}^{n} D^i\,.$$

Then the solution λ *of*

$$\min_{\lambda \in D} \Psi(\lambda) \tag{3.8}$$

satisfies (3.6) *and* (1.14b) *with* $\delta = \partial\Psi/\partial\lambda$.

Proof

It follows from the definition of D and the fact that $0 \in D$ that for any closed and bounded set Q such that $0 \in Q$, $Q \cap D$ is non-empty, closed and bounded. The set

$$L(\alpha) = \{\lambda \mid \lambda \in D, \Psi(\lambda) \leqq \alpha\}$$

is called a level set, ORTEGA and RHEINBOLDT [17, 4.2.1]. Choose $\alpha \geqq 0$. Since Ψ is continuous, G^TG is positive definite and $\Psi(0) = 0$, the set $Q = \{\lambda \mid \Psi(\lambda) \leqq \alpha\}$ is closed and bounded and $0 \in Q$. Hence, L has the same properties. By [17, 4.2.2], there is a $\lambda^* \in D$ minimizing Ψ on D.

Apply lemma 1 to (3.8). In particular, there are no directions $p_i \in R^{n+s}$, $i = 1, 2, \dots, n$, such that $\lambda^* + \alpha p_i \in D$, $\alpha \neq 0$, where

$$p_{ij} = \begin{cases} 0, & j \neq i, \quad j \neq n+i, \\ p_N, & j = i, \\ p_F, & j = n+i, \end{cases}$$

$$(\nabla\Psi(\lambda^*))^T p_i < 0 .$$

Therefore, it is sufficient to consider the one-contact problem discussed above. We found that if $b_i(\lambda^*) \leqq 0$ then

$$|\lambda_{Fi}^*| < \mu\lambda_{Ni}^* \Rightarrow \delta_{Ni} = \delta_{Fi} = 0 , \qquad |\lambda_{Fi}^*| = \mu\lambda_{Ni}^* \Rightarrow \lambda_{Fi}^*\delta_{Fi} \leqq 0$$

and if $b_i(\lambda^*) > 0$ then

$$\lambda_{Ni}^* = \lambda_{Fi}^* = 0 \quad \text{and} \quad \delta_{Ni} > 0 .$$

For $s < i \leqq n$, it is easily shown by means of the necessary conditions that

$$\lambda_{Ni}^* \geqq 0 , \qquad \delta_{Ni} \geqq 0 , \qquad \lambda_{Ni}^*\delta_{Ni} = 0 .$$ ■

Unfortunately, D is not convex, which is realized e.g. by considering a 3×3 system with one friction and two normal multipliers. Therefore, we cannot claim that the solution of (3.8) is unique.

Next we shall derive sufficient conditions for the general problem to ensure existence of a solution. Pairs $(\lambda_{Ni}, \delta_{Ni})$ and $(\lambda_{Fi}, \delta_{Fi})$ satisfying (1.14) are determined by an algorithm which terminates at the solution if the sufficient conditions are fulfilled. The following special case demonstrates the relation of the friction problem to linear complementarity problems (LCP). Suppose that $J'_{F0} = \emptyset$. Then $\lambda_{Fi} = \mu s_i\lambda_{Ni}$, $i \in J_F$, $s_i = +1$ or -1, and (1.14) simplifies to

$$G_1^T(G_1 + HU)\lambda_N + b = \delta_N \geqq 0 , \quad \lambda_N \geqq 0 , \quad \lambda_N^T\delta_N = 0 , \quad \delta_N^T = (\delta_{N1}, \dots, \delta_{Nn}) , \quad b = \dot{G}_1^T\dot{q} + G_1^Tf \tag{3.9}$$

where $\lambda_N \geqq 0$ is short for $\lambda_{Ni} \geqq 0$, $i = 1, 2, \dots, n$. Observe that $\lambda_N^T\delta_N = 0$ is equivalent to $\lambda_{Ni}^T\delta_{Ni} = 0$, $i = 1, 2, \dots, n$. (3.9) is an LCP. In order to state a result on the solution of (3.9) a definition is necessary.

Definition

A square matrix A *belongs to the class of* P*-matrices if it has positive principal minors.*

It is proved in COTTLE and DANTZIG [5] that if $G_1^T(G_1 + HU)$ is a P-matrix, then there is a solution to (3.9) for all b. If $\mu = 0$ then $HU = 0$ and $G_1^TG_1$ is symmetric and at least positive semi-definite. If there is a c such that $b = G_1^Tc$, then a solution exists, INGLETON [10].

The principal pivoting method of COTTLE and DANTZIG [5], which is directly applicable to systems with only normal multipliers λ_{Ni} as in (3.9) is modified in order to also allow systems with friction multipliers λ_{Fi} obeying Coulomb's law. We seek the solution to (cf. (1.14) and (3.9))

$$G_1^T(G_1 + HU)\lambda_1 + b = \delta_1 , \tag{3.10a}$$

$$\lambda_{Ni} \geqq 0 , \qquad \delta_{Ni} \geqq 0 , \qquad \lambda_{Ni}^T\delta_{Ni} = 0 , \qquad i \in J'_N , \tag{3.10b}$$

$$-\mu\lambda_{Ni} \leqq \lambda_{Fi} \leqq \mu\lambda_{Ni} , \qquad i \in J'_{F0} ,$$

$$\text{if} \quad |\lambda_{Fi}| < \mu\lambda_{Ni} \quad \text{then} \quad \delta_{Fi} = 0 , \qquad \text{if} \quad |\lambda_{Fi}| = \mu\lambda_{Ni} \quad \text{then} \quad \lambda_{Fi}\delta_{Fi} \leqq 0 . \tag{3.10c}$$

In accordance with the terminology of mathematical programming, a pair $(\lambda_{Ni}, \delta_{Ni})$ is feasible if it satisfies the relations in (3.10b), otherwise it is non-feasible. A pair $(\lambda_{Fi}, \delta_{Fi})$ is feasible only if $(\lambda_{Ni}, \delta_{Ni})$ is and $(\lambda_{Fi}, \delta_{Fi})$ satisfies (3.10c). The algorithm for sloving (3.10) is based on principal pivoting and the selection of a driving variable, two concepts which will now be explained.

Let $y, x, c \in R^n$, $A \in R^{n \times n}$ and

$$y = Ax + c . \tag{3.11}$$

The variables in x are termed fixed and those in y free, except for one in x, x_d, the driving variable, which is changed continuously in order for x_d and y_d to satisfy relations such as (3.10b) and (3.10c). Suppose that is possible to make an interchange so that one of the x-variables $x_j \neq x_d$ is expressed as a unique function of y_j and the rest of the fixed variables. Define

$$\bar{y}^T = (y_1, y_2, \dots, y_{j-1}, x_j, y_{j+1}, \dots, y_n), \qquad \bar{x}^T = (x_1, x_2, \dots, x_{j-1}, y_j, x_{j+1}, \dots, x_n).$$

Then there are $\bar{A} \in R^{n\times n}$ and $\bar{c} \in R^n$ such that

$$\bar{y} = \bar{A}\bar{x} + \bar{c}. \tag{3.12}$$

$\bar{A}$ is a principal pivot transform of A. If $\lambda_{Ni} > 0$, $\delta_{Ni} > 0$, $|\lambda_{Fi}| < \mu\lambda_{Ni}$ or $\delta_{Fi} \neq 0$ in our problem (3.10), they are all free variables and the other variable in the pair is fixed, $\delta_{Ni} = 0$, $\lambda_{Ni} = 0$, $\delta_{Fi} = 0$ or $|\lambda_{Fi}| = \mu\lambda_{Ni}$. In the algorithm δ_{Nd} or δ_{Fd} is the driving variable. When a free variable y_j in a feasible pair reaches its extremal value, $\lambda_{Ni} = 0$, $\delta_{Ni} = 0$, $|\lambda_{Fi}| = \mu\lambda_{Ni}$ or $\delta_{Fi} = 0$, due to the change of the driving variable x_d, the system of equations (3.11) is transformed by means of principal pivoting so that x_j becomes the new free variable and y_j is fixed. Then the change of x_d can be continued, while the feasibility of all the other pairs (x_i, y_i), $i \neq d$, is retained. If λ_{Fj} is fixed and λ_{Nj} is free then the two free variables δ_{Fj} and λ_{Nj} are functions of only δ_{Nj} and the fixed variables from the other points of contact $i \neq j$. Hence, the number of columns of $\bar{A}$ is reduced by one compared to the number of A when $|\lambda_{Fj}|$ attains its upper bound $\mu\lambda_{Nj}$.

The algorithm is:

0. Make an initial guess

$$\lambda_1 = (G_1^T(G_1 + HU))^{-1}\,\delta_1 - (G_1^T(G_1 + HU))^{-1}\,b \tag{3.13}$$

with $\delta_1 = 0$.

1. If there is a non-feasible pair $(\lambda_{Ni}, \delta_{Ni})$, i.e. $\lambda_{Ni} < 0$ and $\delta_{Ni} = 0$, then take δ_{Ni} as driving variable and increase δ_{Ni} until the pair is feasible, while maintaining the feasibility of all the previously feasible pairs $(\lambda_{Nj}, \delta_{Nj})$ and $(\lambda_{Fk}, \delta_{Fk})$. When all pairs $(\lambda_{Ni}, \delta_{Ni})$ are feasible then proceed with step 2.
2. If there is a non-feasible pair $(\lambda_{Fi}, \delta_{Fi})$, i.e. $|\lambda_{Fi}| > \mu\lambda_{Ni}$ and $\delta_{Fi} = 0$, then if $\lambda_{Fi} < 0$ increase or if $\lambda_{Fi} > 0$ decrease δ_{Fi} until the pair is feasible, while maintaining the feasibility of already feasible pairs. When all pairs $(\lambda_{Fi}, \delta_{Fi})$ are feasible, then the whole solution, λ_1 and δ_1, is feasible.

Lemma 2 provides sufficient conditions on $G_1^T(G_1 + HU)$ to ensure that a pair can be made feasible in the way indicated above without destroying the feasibility of the other pairs. The subsequent theorem proves that the algorithm converges globally for all b. The case of degeneracy, where two or more variables from different points of contact, $(\lambda_{Ni}, \delta_{Ni}, \lambda_{Fi}, \delta_{Fi})$ and $(\lambda_{Nj}, \delta_{Nj}, \lambda_{Fj}, \delta_{Fj})$, attain their extrema simultaneously is not taken into account in the analysis.

Lemma 2

Let J_{F1} be any subset (including $\emptyset$) of J'_{F0} and $J_{F2} = J'_{F0} \setminus J_{F1}$. Define $G_2 = (G_N, g_{Fi_1}, \dots, g_{Fi_l})$, $\lambda_2^T = (\lambda_N^T, \lambda_{Fi_1}, \dots, \lambda_{Fi_l})$, $i_j \in J_{F1}$ and $H_2 = (H, g_{Fi_1}, \dots, g_{Fi_s})$, $i_j \in J_{F2}$. Let U_2 be determined by

$$\sum_{i\in J'_{FW}} g_{Fi}s_i\mu\lambda_{Ni} + \sum_{i\in J_{F2}} g_{Fi}s_i\mu\lambda_{Ni} = H_2U_2\lambda_2$$

(cf. (1.11)). If $G_2^T(G_2 + H_2U_2)$ is a P-matrix for any set J_{F1} with any possible choice of $s_i = \pm 1$, $i \in J_{F2}$, then a non-feasible pair in the solution of (3.10) can be made feasible by a continuous increase or decrease of δ_{Ni} or δ_{Fi} while the pairs already feasible remain so.

Proof

The constructive proof is long and tedious and is therefore only outlined here. The full details of the proof, and thus the algorithm, are given in LÖTSTEDT [14]. It is divided into two parts: Firstly, it is shown that a non-feasible pair can be made feasible by changing δ_{Ni} or δ_{Fi}. Secondly, it is demonstrated how the feasibility of a pair is maintained while δ_{Ni} or δ_{Fi} of another pair is changed. All the possible cases of variables reaching their extrema are examined. The proof relies on a result by TUCKER [21] asserting that the principal pivot transforms of P-matrices, also belong to the class P, see also COTTLE and DANTZIG [5]. ■

Theorem 2

If the conditions on G_1 and H in lemma 2 are satisfied, then for every b the algorithm terminates in a solution of (3.10).

Proof

The proof can be carried through using the same kind of arguments as those in the proofs of theorems 11 and 12 in COTTLE and DANTZIG [5]. ■

The linear complementarity problem (3.9) is contained in (3.10) and when $J'_{F0} = \emptyset$, theorem 2 is equivalent to the results from COTTLE and DANTZIG [5] for (3.9) cited above. In example 1, $a = G_1^T(G_1 + HU) < 0$ and theorem 2 does not guarantee the existence of a solution for all b. However, the condition in lemma 2 is not necessary. The algorithm produces a solution to examples where G_1 and H do not have the properties required in lemma 2.

Somtime a normal multiplier term $g_{Ni}\lambda_{Ni}$ is introduced into the equations of motion (1.12a) so that $q(t)$ fulfils a bilateral contraint $\phi_i = 0$ g JO $g_{Ni}^T\dot{q} = 0$ for $t \geq 0$. Then $\delta_{Ni} = 0$ and λ_{Ni} is allowed to take any sign, PÉRÈS [19]. The modulus of λ_{Fi} is there restricted by $|\lambda_{Fi}| \leq \mu\,|\lambda_{Ni}|$. If the sets D_1, D_2 and D_3 in the one-contact example (3.5) are replaced by

$$D_1' = \{\lambda_N, \lambda_F \mid |\lambda_F| \leqq \mu \, |\lambda_N|\} \,, \qquad D_2' = \{\lambda_N, \lambda_F \mid u\lambda_N + v\lambda_F + x = 0\} \,,$$

then the solution to the bilateral one-contact problem will render (3.4) a minimum over $D_1' \cap D_2'$. Theorem 1 is valid also when bilateral contraints are included. After minor modifications in the algorithm and the proof of lemma 2, the above mentioned relations to be satisfied by λ_{Ni} and λ_{Fi} can be incorporated in (3.10). The sufficient condition in lemma 2 remains the same.

In an edge-to-edge contact between two bodies (fig. 5) the reaction forces can be represented by two normal forces at each end of the line of contact and one friction force. Let the corresponding terms in the equations of motion from contact j be $g_{Nj}^1\lambda_{Nj}^1$, $g_{Nj}^2\lambda_{Nj}^2$ and $g_{Fj}\lambda_{Fj}$. The multipliers and the associated relative accelerations δ_{Nj}^1, δ_{Nj}^2 and δ_{Fj} satisfy

$$\lambda_{Nj}^1 \geqq 0 \,, \quad \delta_{Nj}^1 \geqq 0 \,, \quad \lambda_{Nj}^1\delta_{Nj}^1 = 0; \quad \lambda_{Nj}^2 \geqq 0 \,, \quad \delta_{Nj}^2 \geqq 0 \,, \quad \lambda_{Nj}^2\delta_{Nj}^2 = 0;$$

$$|\lambda_{Fj}| \leqq \mu(\lambda_{Nj}^1 + \lambda_{Nj}^2) \,, \quad (\mu(\lambda_{Nj}^1 + \lambda_{Nj}^2) - |\lambda_{Fj}|) \, \delta_{Fj} = 0 \,, \quad \lambda_{Fj}\delta_{Fj} \leqq 0 \,. \tag{3.14}$$

g_{Nj}^1, g_{Nj}^2 and g_{Fj} are linearly independent. There is a modification of the sets D_1, D_2 and D_3 in (3.5) such that the minimum of (3.4) satisfies (3.14). It can be shown that theorem 1 holds also for systems with edge-to-edge contacts. The redefinition of U in (1.11) to take (3.14) into account is obvious. The definitions of the sets J_{F0}', J_{F1} and J_{F2} are not altered in lemma 2. Using the same technique as in the proof of the lemma and theorem 2, it can be proved that the same conditions on $G_2^T(G_2 + H_2U_2)$ as in the lemma guarantee the existence of a solution also when the relations (3.14) are to be satisfied for some contacts.

4. Sufficient conditions for uniqueness

A matrix A in the class P has the following characterization given by FIEDLER and PTÁK [9]: For every vector $x \neq 0$, there exists an index i such that $x_i(Ax)_i > 0$. This property will be employed to derive a sufficient condition for the solution of (3.10) to be unique. Using the assertion above, MURTY [16] proves that the solution of (3.9) is unique if and only if $G_1^T(G_1 + HU)$ is a P-matrix.

To state our result we need some notation. For each λ_{Fi} there is a μ_i, $-\mu \leqq \mu_i \leqq \mu$, such that $\lambda_{Fi} = \mu_i\lambda_{Ni}$. If $i \in J_{FW}'$ then $|\mu_i| = \mu$. It is convenient to introduce

$$G^* = (g_{N1} + g_{F1}\mu_1, g_{N2} + g_{F2}\mu_2, \dots, g_{Nr} + g_{Fr}\mu_r, g_{N,(r+1)}, \dots, g_{Nn})$$

so that the resulting contact forces $G\lambda$ in (1.13a) can be written

$$G\lambda = \sum_{i=1}^{n} g_{Ni}\lambda_{Ni} + \sum_{i=1}^{r} g_{Fi}\lambda_{Fi} = \sum_{i=1}^{r} (g_{Ni} + g_{Fi}\mu_i) \lambda_{Ni} + \sum_{i=r+1}^{n} g_{Ni}\lambda_{Ni} = G^*\lambda_N = (G_1 + HU) \lambda_1 \,. \tag{4.1}$$

G_F was defined in (3.1).

Theorem 3

If $G_1^T(G^*, G_F)$ is a P-matrix for all μ_i, $-\mu \leqq \mu_i \leqq \mu$, $i \in J_{F0}'$, then (3.10) *has one and only one solution for every b.*

Proof

To begin with, let $B = G_1^T(G_1 + HU)$. In (3.10) $B = G_1^T(G^*, G_F)$ with all $\mu_i = 0$, $i \in J_{F0}'$. Hence, B is a P-matrix and by theorem 2 a solution exists.

Suppose that there is a b such that (3.10) has two different solutions $\lambda_N^1, \lambda_F^1, \delta_N^1, \delta_F^1$ and $\lambda_N^2, \lambda_F^2, \delta_N^2, \delta_F^2$ where for $i = 1, 2$,

$$\lambda_F^i = (\lambda_{Fj_1}^i, \lambda_{Fj_2}^i, \dots, \lambda_{Fj_s}^i) \,, \qquad j_k \in J_{F0}' \,.$$

Define U^i such that $\lambda_F^i = U^i\lambda_N^i$, $U^i \in R^{s\times n}$. Each row of U^i has only one potential non-zero element which lies between $-\mu$ and μ. The difference between the solutions fulfils

$$B\begin{pmatrix} \lambda_N^1 - \lambda_N^2 \\ U^1\lambda_N^1 - U^2\lambda_N^2 \end{pmatrix} = \begin{pmatrix} \delta_N^1 - \delta_N^2 \\ \delta_F^1 - \delta_F^2 \end{pmatrix}. \tag{4.2}$$

Insert the identity

$$U^1\lambda_N^1 - U^2\lambda_N^2 = U^1(\lambda_N^1 - \lambda_N^2) + (U^1 - U^2) \lambda_N^2$$

into (4.2) yielding

$$B\begin{pmatrix} I_n & 0 \\ U^1 & I_s \end{pmatrix}\begin{pmatrix} \lambda_N^1 - \lambda_N^2 \\ (U^1 - U^2) \lambda_N^2 \end{pmatrix} = \begin{pmatrix} \delta_N^1 - \delta_N^2 \\ \delta_F^1 - \delta_F^2 \end{pmatrix}, \tag{4.3}$$

where I_k is the $k \times k$ identity matrix.

By virtue of (4.1),

$$A = B\begin{pmatrix} I_n & 0 \\ U^1 & I_s \end{pmatrix} = G_1^T(G^*, G_F) \,,$$

where μ_i in G^* are from U^1 and U. Thus, A is a P-matrix. The two solutions satisfy

$$(\lambda_{Ni}^1 - \lambda_{Ni}^2)(\delta_{Ni}^1 - \delta_{Ni}^2) = -\lambda_{Ni}^1\delta_{Ni}^2 - \lambda_{Ni}^2\delta_{Ni}^1 \leqq 0 .$$

The non-zero elements of U^1 and U^2 have identical positions in the matrices but may have different signs and moduli. For each row j in U^i, i.e. each λ_{Fj}^i,

$$c_j = (U_{jk}^1 - U_{jk}^2)\lambda_{Nk}^2(\delta_{Fj}^1 - \delta_{Fj}^2) = (U_{jk}^1 - U_{jk}^2)\lambda_{Nk}^2\delta_{Fj}^1 + (U_{jk}^2 - U_{jk}^1)\lambda_{Nk}^2\delta_{Fj}^2 .$$

$\lambda_{Nk}^2 \geqq 0$ and if $\delta_{Fj}^1 > 0$ then $U_{jk}^1 = -\mu \leqq U_{jk}^2$ and $(U_{jk}^1 - U_{jk}^2)\lambda_{Nk}^2\delta_{Fj}^1 \leqq 0$. If $\delta_{Fj}^1 < 0$ then $U_{jk}^1 = \mu \geqq U_{jk}^2$ and the same holds true. Similarly, $(U_{jk}^2 - U_{jk}^1)\lambda_{Nk}^2\delta_{Fj}^2 \leqq 0$ and $c_j \leqq 0$.

The conclusion is that, introducing

$$x = \begin{pmatrix} \lambda_N^1 - \lambda_N^2 \\ (U^1 - U^2)\lambda_N^2 \end{pmatrix}, \qquad y = \begin{pmatrix} \delta_N^1 - \delta_N^2 \\ \delta_F^1 - \delta_F^2 \end{pmatrix}, \qquad Ax = y ,$$

we have

$$x_i(Ax)_i \leqq 0 , \qquad \forall i ,$$

which is contradictory to the property of matrices in the class P stated in the beginning of the section unless $x = 0$. It follows that $\lambda_N^1 = \lambda_N^2$ and $(U^1 - U^2)\lambda_N^2 = 0$. If $\lambda_{Nj}^2 = \lambda_{Nj}^1 = 0$ then the difference $U_{ij}^1 - U_{ij}^2$ is unimportant, otherwise $U_{ij}^1 = U_{ij}^2$ implying that $\lambda_F^1 = \lambda_F^2$. ■

In example 2 where three different solutions were possible, $G_1 = (g_N, g_F)$ and $G^* = g_N + g_F\mu^*$. If $\mu^* = \mu$ then $g_N^T(g_N + g_F\mu^*) < 0$ and $G_1^T(G^*, g_F)$ is evidently not a P-matrix for $\mu^* \in [-\mu, \mu]$.

If G_1 has linearly independent columns then $G_1^TG_1$ is in the class P. Let $p(A)$ denote an arbitrary principal minor of A. Since $p(A)$ is defined as a sum of products of the elements of A, $p(A + \varepsilon B) = p(A) + O(\varepsilon)$. For a sufficiently small μ, both $p\big(G_1^T(G_1 + HU)\big) = p(G_1^T G_1) + O(\mu)$ and $p\big(G^T(G^*, G_F)\big) = p(G_1^TG_1) + O(\mu)$ are positive and the conditions in lemma 2 and theorem 3 are fulfilled.

5. On the solution in a time-interval

In sections 2, 3 and 4 we have discussed the problem of dropping constraints from J_N and J_{F0} at t_0. We shall here turn to the question of adding constraints and to the solution of (1.12) in a time-interval.

It is well-known that when a previously positive ϕ_k vanishes at t_1, e.g. after a collision between two bodies, $\dot{q}$ is in general discontinuous at t_1. Let the superscripts 1 and 2 on a variable denote its value immediately before and immediately after the discontinuity. At t_1 $\dot{q}^2$ satisfies the jump condition

$$M(q^2)(\dot{q}^2 - \dot{q}^1) = G(q^2)\,\Lambda . \tag{5.1}$$

The components of Λ are Lagrange multipliers proportional to impulses in the mechanical system. Various conditions on Λ_i and $(g_i^T\dot{q})^2$ are possible, see BOULIGAND [3], KILMISTER and REEVE [12] and PÉRÈS [19]. They are all similar to those on λ_i and δ_i in (1.12). J_N at $t_1 - 0$ is augmented by k to form J_N at $t_1 + 0$.

Suppose that an index k is dropped from J_{FW} and added to J_{F0} at t_1. Then $g_{Fk}^T\dot{q}$ reaches zero at t_1 and in general δ_{Fk}, λ and $\ddot{q}$ are discontinuous. This is illustrated by a box sliding on a slope, whose downward motion stops due to friction. There is usually no difficulty in determining when and which constraints to add to J_N and J_{F0}, since the decisions are based on algebraic functions switching sign.

In intervals (t_0, t_1) between the updates of J_N, J_{F0} and J_{FW}, $\delta_{Ni} = 0$, $i \in J_N$, and $\delta_{Fi} = 0$, $i \in J_{F0}$. Hence, by (1.12b) the governing system of equation in (t_0, t_1) is

$$\ddot{q} = f + (G_1 + HU)\lambda_1 , \qquad \delta_1 = G_1^T\ddot{q} + \dot{G}_1^T\dot{q} = 0 . \tag{5.2a), (5.2b}$$

Replace $\ddot{q}$ in (5.2b) by $\ddot{q}$ from (5.2a) and suppose that $G_1^T(G_1 + HU)$ is non-singular. Then solve the resulting equation for λ_1:

$$\lambda_1 = -\big(G_1^T(G_1 + HU)\big)^{-1}(G_1^Tf + \dot{G}_1^T\dot{q}) . \tag{5.3}$$

After inserting λ_1 from (5.3) into (5.2a) the system of differential equations to be solved in (t_0, t_1) is

$$\ddot{q} = \big(I - (G_1 + HU)(G_1^T(G_1 + HU))^{-1}G_1^T\big)f - (G_1 + HU)\big(G_1^T(G_1 + HU)\big)^{-1}\dot{G}_1^T\dot{q} . \tag{5.4}$$

Assume that the initial conditions $q_0 = q(0)$ are such that $\phi_i(q_0) \geqq 0$, $\forall_i$. q_0 together with $\dot{q}_0 = \dot{q}(0)$ determine J_N', J_{F0}' and J_{FW}' at $t = 0$. Then use the procedure described in section 3 to compute $\lambda_1(0)$ (if possible). Let us disregard degenerate systems and assume that for all pairs $(\lambda_{Ni}, \delta_{Ni})$ either $\lambda_{Ni} > 0$ or $\delta_{Ni} > 0$ and for all pairs $(\lambda_{Fi}, \delta_{Fi})$ either $\mu\lambda_{Ni} - |\lambda_{Fi}| > 0$ or $\delta_{Fi} \neq 0$ at $t = 0$. Introduce new sets defined by $\lambda_1(0)$ and $\delta_1(0)$:

$$\begin{aligned} J_N'' &= J_N' \cap \{i \mid \lambda_{Ni}(0) > 0\} , \qquad J_{F0}'' = J_{F0}' \cap J_N'' \cap \{i \mid \mu\lambda_{Ni}(0) - |\lambda_{Fi}(0)| > 0\} , \\ J_{FW}'' &= (J_{FW}' \cup \{i \mid \delta_{Fi}(0) \neq 0\}) \cap J_N'' . \end{aligned} \tag{5.5}$$

If $\lambda_1(t)$ is continuous and if there are no $\phi_i(0) > 0$ or $g_{Fi}^T\dot{q}(0) \neq 0$ vanishing in $(0, t')$, then $J_N = J_N''$, $J_{F0} = J_{F0}''$ and $J_{FW} = J_{FW}''$ are constant in $(0, t_0)$ for some t_0, $0 < t_0 \leqq t'$. The system of differential equations to be solved in $(0, t_0)$ is (5.4), where G_1 and H are determined by J_N, J_{F0} and J_{FW} in $(0, t_0)$. When λ_{Ni}, $i \in J_N$, ϕ_i, $i \notin J_N$, $\mu\lambda_{Ni} - |\lambda_{Fi}|$, $i \in J_{F0}$, or $g_{Fi}^T\dot{q}$, $i \in J_{FW}$, vanishes at t_0, the sets J_N'', J_{F0}'' and J_{FW}'' and G_1, H are updated as in (5.5) using $\lambda_1(t_0)$, $\delta_1(t_0)$. If ϕ_i has reached zero then a new velocity $\dot{q}$ at $t_0 + 0$ must first be computed using (5.1). The updated sets and matrices define a new system (5.4) to be integrated in (t_0, t_1) for some t_1, $t_0 < t_1$. The theorems 2 and 3 are combined with the classical uniqueness result for ordinary differential equations in the concluding theorem.

Theorem 4

Assume that G_1 and H satisfy the conditions in lemma 2 at $t = t_0$, that G_1, H, f and $\partial G_1/\partial q$ are bounded and Lipschitz continuous in the neighbourhood of $q(t_0)$, $\dot{q}(t_0)$ and t_0, that $\lambda_1(t_0)$ satisfies the non-degeneracy condition

$$\text{if } \lambda_{Ni} = 0 \text{ then } \delta_{Ni} > 0\,, \quad i \in J'_N \text{ at } t_0\,, \quad \text{if } \mu\lambda_{Ni} - |\lambda_{Fi}| = 0 \text{ then } \delta_{Fi} \neq 0\,, \quad i \in J'_{F0} \text{ at } t_0$$

and that no $\phi_i > 0$ or $g_{Fi}^T \dot{q} \neq 0$ vanishes in (t_0, t'), $t' > t_0$. Then there exists a solution to (1.12) in $[t_0, t_1)$ for some t_1, $t_0 < t_1 \leqq t'$. Furthermore, if the condition in theorem 3 on G_1 and H is satisfied at t_0, then the solution is unique.

Proof

$\dot{G}_1 = \dfrac{\partial G_1}{\partial q}\,\dot{q}$ is Lipschitz continuous. By theorem 2 a solution $\lambda_1(t_0)$ exists. Let $\lambda_1(t_0)$ and J'_N, J'_{F0} and J'_{FW} at t_0 determine the sets in (5.5). In the new matrices G_1 and $G_1 + HU$ a few columns may have been dropped from the old ones defined by J'_N and J'_{F0} at t_0. The new $G_1^T(G_1 + HU)$ is a principal submatrix of one of the previous $G_2^T(G_2 + H_2U_2)$ in lemma 2. Since $G_2^T(G_2 + H_2U_2)$ is a P-matrix, so is $G_1^T(G_1 + HU)$. Thus, $\big(G_1^T(G_1 + HU)\big)^{-1}$ exists and by Cramer's rule for inversion of matrices it is easy to show that $\big(G_1^T(G_1 + HU)\big)^{-1}$ also is Lipschitz continuous. In view of (5.3), $\lambda_1(t)$ is continuous in t and the governing system (5.4) in (t_0, t_1) is determined according to the discussion preceding the theorem. By the Lipschitz continuity of the right hand side of (5.4) it follows from CODDINGTON and LEVINSON [4] that given the sets J''_N and J''_{F0}, the solution $q(t)$, $\dot{q}(t)$ is exists and is unique in the neighbourhood of t_0. If the conditions of theorem 3 are fulfilled, then J''_N and J''_{F0} are unique. Hence, $G_1^T(G_1 + HU)$ is unique and apparently also $\lambda_1(t)$, $\ddot{q}(t)$, $\dot{q}(t)$ and $q(t)$. ■

Acknowledgement

This paper is extracted from the author's Ph. D-thesis. During the preparation of the thesis the encouragement and advice by Professor G. DAHLQUIST has been invaluable. Financial support has been obtained from The Swedish Institute of Applied Mathematics and The National Swedish Board for Technical Dvelopment.

References

1 BÉGHIN, H., Sur certain problèmes de frottement, Nouvelles Annales, 5e série, 2 (1923–24), p. 305–312.
2 BÉGHIN, H., Sur l'indétermination de certain problèmes de frottement, Nouvelles Annales, 5e série, 3 (1924–25), p. 343–347.
3 BOULIGAND, G., Mécanique rationnelle, 5e ed., Vuibert, Paris, 1954.
4 CODDINGTON, E. A.; LEVINSON, N., Theory of ordinary differential equations, McGraw-Hill, New York, 1955.
5 COTTLE, R. W.; DANTZIG, G. B., Complementary pivot theory of mathematical programming, Lin. Alg. Appl., 1 (1968), p. 103 to 125.
6 CUNDALL, P. A.; STARK, O. D. L., A discrete model for granular assemblies, Géotechnique, 1 (1979), p. 47–65.
7 DELASSUS, Ét., Considérations sur le frottement de glissement, Nouv. Ann. de Mathématique, 4e série, 20 (1920), p. 485–496.
8 DUVAUT, G.; LIONS J. L., Les inéquations en mécanique et en physique, Dunod, Paris, 1972.
9 FIEDLER, M.; PTÁK, V., On matrices with non-positive off-diagonal elements and positive principal minors, Czech. Math. J., 12 (1962), p. 382–400.
10 INGLETON, A. W., A problem in linear inequalities, Proc. London Math. Soc. (3), 16 (1966), p. 519–536.
11 KARAMARDIAN, S., The complementarity problem, Math. Prog., 2 (1972), p. 107–129.
12 KILMISTER, C. W.; REEVE, J. E., Rational mechanics, Longmans, London, 1966.
13 KLEIN, F., Zu Painlevés Kritik der Coulombschen Reibungsgesetze, Z. f. Math. und Phys., 58 (1909), S. 186–191.
14 LÖTSTEDT, P., Analysis of some difficulties encountered in the simulation of mechanical systems with constraints, Dept Numerical Analysis and Computing Science, Royal Inst. of Techn., TRITA-NA-7914, Stockholm, 1979.
15 LÖTSTEDT, P.; DAHLQUIST, G., Interactive simulation of the progressive collapse of a house, In: A. W. BENNET, R. VICHNEVETSKY (Ed), Num. meth. for differential equations and simulation, North-Holland, New York, 1978, p. 135–138.
16 MURTY, K., On the number of solutions to the complementarity problem and spanning properties of complementarity cones Lin. Alg. Appl., 5 (1972), p. 65–108.
17 ORTEGA, J. M.; RHEINBOLDT, W. C., Iterative solution of non-linear equations in several variables, Academic Press, New York, 1970.
18 PAINLEVÉ, P., Sur les lois du frottement de glissement, C.R. Acad. Sci., 121 (1895), p. 112–115, 140 (1905), p. 702–707, 141 (1905), p. 401–405 et 546–552.
19 PÉRÈS, J., Mécanique générale, Masson & Cie, Paris, 1953.
20 POWELL, M. J. D., Introduction to constrained optimization, In: P. E. GILL, W. MURRAY (Ed.), Numerical methods for constrained optimization, Academic Press, London, 1974, p. 1–28.
21 TUCKER, A. W., Principal pivotal transforms of square matrices, SIAM Review, 5 (1963), p. 305.

Eingereicht am 21. 10. 1980

Anschrift: Dr. PER LÖTSTEDT, Department of Numerical Analysis and Computing Science, Royal Institute of Technology, S-10044 Stockholm 70, Sweden

ZAMM 61, 617 – 627 (1981)

G. STOYAN / H. BAUMERT

Parameter Identification in Transverse Mixing Models of Rivers — an Inverse Problem for a Parabolic Equation

The basic equation of transverse mixing in rivers and streams is given. A simple analytical solution (Gaussian distribution) is derived, firstly, assuming uniform transverse distributions of longitudinal flow velocity, depth, and transverse mixing coefficient. Using a modified linear regression technique, the cross sectional average of the transverse mixing coefficient is determined. Then, to get an estimate of the cross sectional distribution of the mixing coefficient, a finite difference scheme was applied in order to solve the mixing equation numerically, and the identification of the mixing coefficient was carried out by a nonlinear least squares technique, combining several minimization methods. The numerical methods described were successfully applied to experimental data of the Elbe river and led to certain crossmixing models. In applications, these models may save measurement expense and serve as scientific tools to predict the influence of large waste water treatment plants on water quality in the longitudinal and transversal coordinate of the river.

Ausgehend von einer parabolischen Gleichung für den Quervermischungsprozeß in Flüssen wird zunächst eine analytische Lösung (Gaußverteilung) hergeleitet, die konstante Werte für die Strömungsgeschwindigkeit, die Tiefe und den Quervermischungskoeffizienten voraussetzt. Durch modifizierte lineare Regression wird der Quervermischungskoeffizient bestimmt. Für den Fall transversal variierender Größen wird ein Differenzenschema zur numerischen Lösung der Gleichung benutzt und der Quervermischungskoeffizient mit Hilfe eines verschiedene Minimierungsmethoden kombinierenden Verfahrens für nichtlineare kleinste Quadrate identifiziert. Die Anwendung der dargestellten Methoden auf Meßdaten der Elbe ergab verschiedene Quervermischungsmodelle, deren Benutzung zur Verringerung von Meßaufwand führen kann, und die geeignet sind, die räumlichen Auswirkungen großer Abwassereinleitungen auf die Wasserqualität vorherzusagen.

Рассмотрено параболическое уравнение для процесса поперечного смешения в реках. Предполагая постоянство скорости течения, глубины и коэффициента поперечного смешения, получено аналитическое решение (распределение Гаусса), с помощью которого, проведя модифицированную линейную регрессию, определен коэффициент поперечного смешения. Для случая величин, зависящих от поперечной координаты, применяется разностная схема и определяется коэффициент, минимизируя подходящий функционал типа суммы нелинейных квадратов с помощью алгоритма, комбинирующего различные методы минимизации. Применение представленного подхода к данным измерения для реки Эльба привело к некоторым моделям поперечного смешения, использование которых сможет сэкономить измерительные затраты и помочь в предсказании пространственного влияния больших источников сточных вод на качество воды.

1. Introduction

In the last years physically founded models of transport, mixing and conversion processes of pollutants in natural rivers were developed, see, e.g., [4], [5], [6], [20], [29], [30]. One of the most actual problems in this particular field of applied mathematical modelling is the identification of the model parameters.

Modelling and forecasting of mixing processes in flowing waters is not only of scientific but also of great practical interest in connection with water quality management and decision making. In particular, pointwise waste water outlets generate strong water quality gradients in the transverse direction which can be indicated in river reaches of more than hundred kilometers length [5], connected with quality problems at the river banks, at the water supply facilities etc.

In the following two approaches are used in order to identify the cross mixing coefficient which is not directly measurable. The first one employs an analytical solution for the constant coefficient case and simple weighted least squares. In the second one, the mixing coefficient is considered to vary in transverse direction. To obtain the coefficient, an inverse problem for the parabolic equation (1) below is formulated and solved by combining a conservative difference scheme and a nonlinear least squares method. Both approaches are applied to stationary point sources which form "natural tracer plumes". In this way cross-mixing models have been obtained without additional injections of artificial tracers.

2. The Basic Equation

Neglecting density effects and the curvature of the river as well as the longitudinal mixing, the mass balance for an infinitely small space element of the river (i.e. for a river slice) is expressed by the following parabolic differential equation [23], [29], [30]:

$$\frac{\partial c}{\partial x} = \frac{\partial}{\partial q}\left(v(x, q)\, D(x, q)\, d^2(x, q)\, \frac{\partial c}{\partial q}\right) + S \tag{1}$$

where

x — longitudinal coordinate of the river
v — local value of the vertically averaged flow velocity
D — local value of the vertically averaged transverse mixing coefficient
d — local depth
S — sinks and sources of the soluted matter (waste water outlets or matter conversion processes)

$q = q(x, y)$ is the so-called partial discharge function, defined by the integral

$$q(x, y) = \int_0^y v(x, y') \cdot d(x, y') \cdot dy' . \tag{2}$$

The reference bank of the river under consideration is defined by $y = 0$ or, equivalently, by $q = 0$. $w = w(x)$ is the local width

$$w(x) = A(x)/\bar{d}(x) ,$$

$A = A(x)$ is the cross sectional area, and $\bar{d}$ is the mean depth at x. $Q(x) = q(x, y = w)$ is the total discharge at x.

Along with the "adiabatic" boundary conditions at the banks

$$\left.\frac{\partial c}{\partial y}\right|_{y=0} = \left.\frac{\partial c}{\partial y}\right|_{y=w} = 0 \tag{3}$$

the equations (1, 2) are accepted and proved as a realistic mathematical model of transverse mixing in large rivers under steady state conditions and with a ratio of width to depth much greater then ten. These conditions are fulfilled in most cases of practical interest.

3. The General Identification Problem

All the quantities of equation (1) can, more or less, explicitely be measured in discrete distances y_j from the reference bank ($j = 1, 2, \ldots, m$), except the transverse mixing coefficient D, which is a derivated physical quantity and not explicitely measurable.

Usually, mixing coefficients such like D are defined locally by the time derivative of the integral quantity σ_y^2, see [17],

$$D = \frac{1}{2} \frac{\partial}{\partial t} \sigma_y^2(t) , \tag{4}$$

where σ_y^2 is the time-dependent variance of a pointwise and instantaneously injected amount of tracer (in y-direction). In the case of a stationary injection holds

$$\frac{\partial}{\partial t} \to v \frac{\partial}{\partial x} .$$

Considering turbulent diffusion as the only mixing mechanism, we get [12], [17]

$$\sigma_y^2(t) = 2 \int_0^t (t - t') \, \Phi_y(t') \, dt' , \tag{5}$$

where Φ_y is the autocovariance function of the Lagrangian velocity fluctuations δv_y in the transverse direction. Hence, for large t, $t \gg \tau_y$ (or large x, $x \gg \bar{v}\tau_y$),

$$D = \tau_y \cdot \overline{(\delta v_y)^2} \qquad (\text{where } [D] = \text{m}^2/\text{s}) . \tag{6}$$

The first factor in (6), τ_y, is the Lagrangian micro-scale, and the second factor the strength of the local transverse turbulence, both depending on the local REYNOLDS number [14], [17],

$$Re = vd/\nu , \tag{7}$$

where ν is the kinematic viscosity.

A polynomial expansion of D with free coefficients γ_l reveals (compare [47])

$$D \sim \sum_l \gamma_l \left(\frac{Re\text{-}Re_{\text{crit}}}{Re_{\text{crit}}}\right)^l \tag{8}$$

Re_{crit} is the critical Reynolds number which has in real world rivers a size of 10^3, while the Reynolds number has a size of 10^4 up to about 10^7 [21].

In real rivers the transverse mixing process is not only driven by turbulent diffusion but by some hydrodynamic dispersion processes (shear effects), too, which originate from secondary and tertiary flows, occuring particularly in the presence of bends, strong spatial bottom variations, and in the case of wind exposition. In such cases the basic equation (1) is either perfectly applicable or it represents a sufficiently accurate approximation of the governing processes. But, as a matter of fact, the transverse mixing coefficient D then gets a much more complex character than in the simple case of pure turbulent diffusion.

Considering the secondary and tertiary flows mentioned above, it becomes very difficult to represent D in a theoretically founded way, i.e. in dependence on the geometric and hydraulic properties of a given river reach. It is easier to determine D from the mixing behaviour of the river in the sense of an input-output analysis, e.g. to calculate D from the concentration field at $x = x_1$, say, resulting from a stationary point source of concentration at $x = 0$, and for different discharge situations. In this way D is made responsible for all those complex processes listed above which are not directly taken into account in (1)—(3).

Spoken mathematically, the inverse problem is posed to determine D from (1)—(3) (the initial condition at $x = 0$ and the point source being represented by S) and from $c(x, q)$ at $x = x_1$.

4. The Gaussian Solution

Approximating the product quantity in (1) by the constant ω, i.e. (see [48])

$$v(x, q)\ D(x, q)\ d^2(x, q) \approx \bar{v}\bar{D}\bar{d}^2 = \omega\ , \tag{9}$$

we find, up to second order of accuracy,

$$\frac{\partial c}{\partial x} = \omega \frac{\partial^2 c}{\partial q^2} + S\ , \tag{10}$$

where

$\bar{v}$ — cross sectional mean velocity
$\bar{D}$ — cross sectional mean transverse mixing coefficient
$\bar{d}$ — cross sectional mean depth.

Equation (10) reveals for a stationary point source at $y = 0$ ($q = 0$) the following Gaussian solution

$$c(x, q) \sim \exp\left(-q^2/4\omega x\right)\ . \tag{11}$$

Considering a homogeneous "background" concentration β we find from (11)

$$\ln\left(c(x, q) - \beta\right) = \text{const} - q^2/4\omega x\ . \tag{12}$$

The boundary conditions have practically no influence on the solution (11).

Hence, the problem is reduced to a linear, weighted regression problem with the following objective function

$$f_I(\bar{D}) = \sum_{j=1}^{m} \left(\tilde{c}(x, q_j) - \beta\right) \cdot \left(\ln\left(\tilde{c}(x, q_j) - \beta\right) - \text{const} + a \cdot aq_j^2\right)^2 \tag{13}$$

where $\tilde{c}(x, q_j)$ are the measured concentrations, and the parameter a is given by

$$a = (4\omega x)^{-1} = (4 \cdot \bar{v} \cdot \bar{D} \cdot \bar{d}^2 \cdot x)^{-1}\ . \tag{14}$$

To prove these and the ideas following later, all necessary data of the Elbe river were collected, i.e. cross sectional distributions of water depth, flow velocity and concentrations of different water soluble chemicals, e.g. of Ca^{++} and Cl^- ions. The hydraulic and geometric data are plotted in fig. 1—3. They represent longitudinally averaged transverse distributions, basing on more then 10 cross sections within $x = 0$ and $x = x_1$.

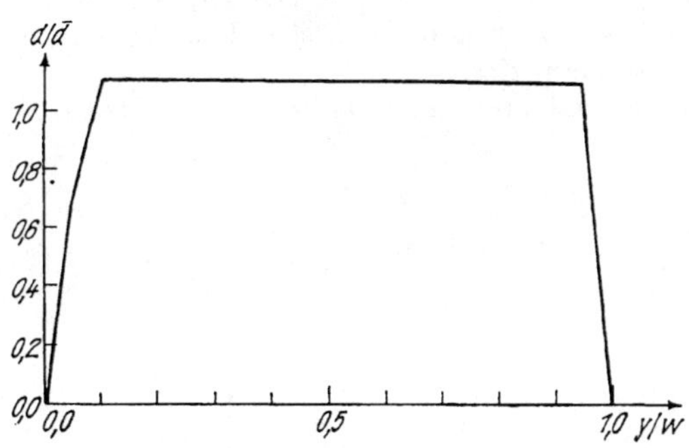

Fig. 1. Averaged and normalized depth function

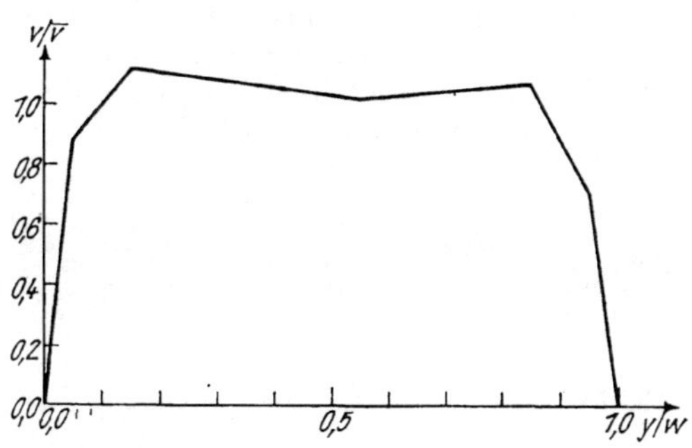

Fig. 2. Averaged and normalized velocity function

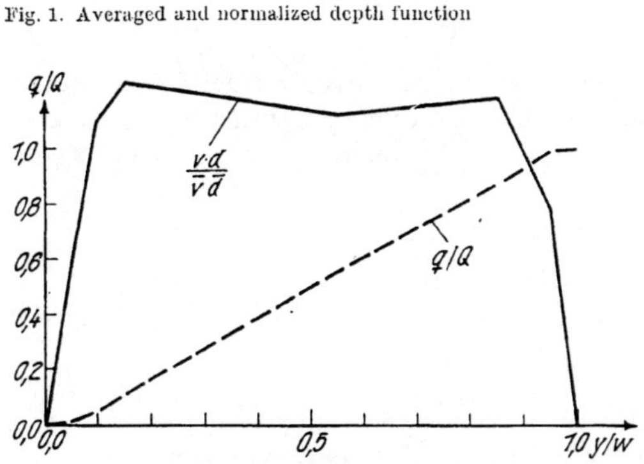

Fig. 3. Averaged and normalized partial discharge function, and product of the averaged and normalized velocity and depth

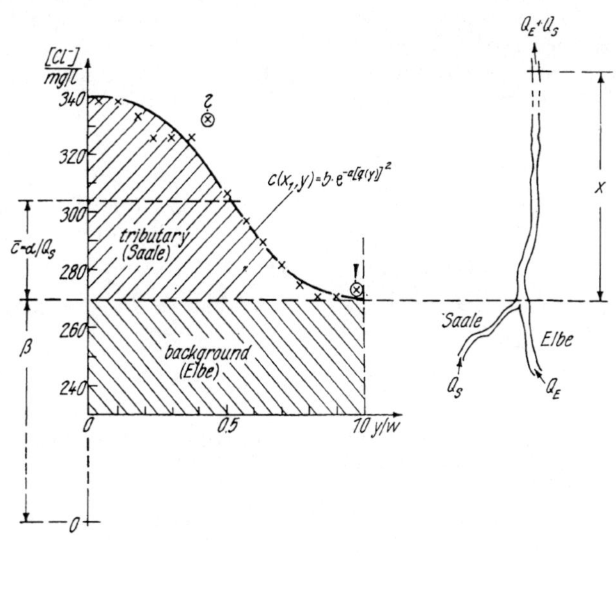

Fig. 4. Definition sketch of the problem and results of the Gaussian model for Cl^- ions

——— Model   × × × Measurements
$|\Delta c|$ = Mean square model error (= 5.1 mg/l ≙ 15%)
$Q_E = 229.8\ m^3/s$, $x_1 = 98500$ m, $D_{mv} = 0.0097\ m^2/s$, $a = 8.4 \cdot 10^{-6}\ m^6/s^2$

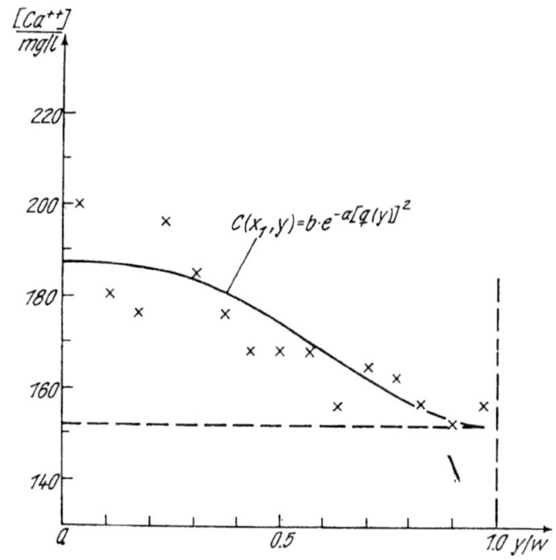

Fig. 5. Results of the Gaussian model for Ca^{++} ions
$Q_E = 229.8\ m^3/s$, $D_{mv} = 0.0097\ m^2/s$, $|\Delta c| = 6.5\ mg/l = 37\%$,
$x_1 = 137\,200\ m$, $\alpha = 6.0 \cdot 10^{-5}\ m^6/s^2$

Assuming (9) and using the RPN algorithm given in [3], p. 236, in order to solve the problem (13), the results shown in fig. 4 and 5 were found by means of a programmable calculator. Obviously, the simple Gaussian model shows no systematic error.

In fig. 4 measured and calculated chloride concentrations are shown. The mean square error of 5.1 mg/l corresponds to a relative model error of approximately 15% ($\overline{\Delta c}/\bar{c} = 5.1/(303-270) = 0.1545$, compare fig. 4). This is practically acceptable. The measurement point which is denoted by a question mark in fig. 4 seems to be explainable by a rough error in chemical analysis. The measurement point denoted by an exclamation mark is caused by one or more small Cl^- sources at the right hand river bank.

In the following analysis the first mentioned concentration value will be ignored. The second mentioned value will be replaced by the background concentration (270 mg/l). Hence the size of the Cl^- data set will be reduced from $m = 15$ to $m = 14$.

The Ca^{++} data set which is plotted in Fig. 5 exhibits a very noisy picture. The relative mean square error is around 37 % and can not be accepted. Therefore, only the Cl^- data set will be used for a further detailed analysis in the case of distributed (i.e. depending on the transverse coordinate y) quantities.

Remark: The quality of all models considered in this article will be evaluated unitarily by means of the relative mean square error criteria $\overline{\Delta c}/\bar{c}$, where

$$\overline{\Delta c} = (f^*/(m-1))^{1/2}\,, \qquad \bar{c} = \frac{1}{m}\sum_{j=1}^{m} c_j - \beta = 33\,.$$

Here f^* is the value of the corresponding functional at the minimizer.

5. Identification of the Distributed Parameter

5.1 Formulation of the inverse problem

In the following we show that it is possible to refine the model (1) in spite of the noise corrupted data. We use the observed spatially distributed functions $v(x, q)$, $d(x, q)$ etc. and replace them by averaged functions $v(y)$, $d(y)$ which are given in Fig. 1 and 2. This "straight channel approximation" is justified by the integrative and averaging character of the mixing process. The connection between y and q gives

$$\frac{\partial}{\partial y} = vd\,\frac{\partial}{\partial q}\,,$$

and the mixing equation (1) is transformed accordingly. Now the unknown distributed parameter $D = D(y)$ will be obtained via solution of an inverse problem in the optimal control setting

$$f_{\rm II}(D) = \min!\,, \qquad D > 0\,, \tag{15}$$

with the objective function $f_{\rm II}$ depending on the distributed parameter D (whereas $f_{\rm I}$ in (13) was a function of the constant $\bar{D}$)

$$f_{\rm II}(D) := \sum_{j=1}^{m} (\tilde{c}_j - c(x_1, y_j, D))^2\,,$$

where $\{\tilde{c}_j\}$ is the set of concentrations of Cl^- measured at $x = x_1$, $y = y_j$, $j = 1, \ldots, m$, and $c(x, y, D)$ is the solution of the parabolic equation

$$v(y)\,d(y)\,\frac{\partial c}{\partial x} = \frac{\partial}{\partial y}\left(D(y)\,d(y)\,\frac{\partial c}{\partial y}\right) + \beta\delta(x) + \alpha\delta(x)\,\delta(y)\,, \qquad 0 < y < w\,, \qquad 0 < x \leqq x_1\,, \tag{16}$$

with boundary conditions

$$\frac{\partial c}{\partial y} = 0\,, \qquad y = 0\,, \qquad y = w\,, \qquad 0 < x \leqq x_1\,.$$

The source term S of (1) is written out in detail here: β denotes the background concentration and α the strength of the point source.

5.2 The numerical solution of the direct problem

In order to solve the inverse problem (15), (16): determine $D(y)$ from given $\{\tilde{c}_j\}$, we need at first a stable, accurate and quick method for the direct problem (16): determine $c(x_1, y_j, D)$ for $D(y)$ given. For this aim, a difference scheme will be used which represents a discrete model of the cross-mixing process. The difference scheme used here has been described in [26]. Applied to equation (16) which contains no separate first order derivative term, the difference scheme is nearly identical to the usual weighted difference scheme (see, e.g., [19], p. 409, p. 431), and takes on the form

$$\frac{1}{2}\,[(vd)_{n+1/2} + (vd)_{n-1/2}]\, c_{x,n} = \frac{1}{h}\left((Dd)_{n+1/2}\, c_{y,n}^{(\sigma_n+1/2)} - (Dd)_{n-1/2}\, c_{y,n-1}^{(\sigma_n-1/2)}\right) \tag{17}$$

for inner points, $n = 1, \ldots, N-1$,

$$\begin{aligned} &\frac{1}{2}\,(vd)_{1/2}\, c_{x,0} = \frac{1}{h}\,(Dd)_{1/2}\, c_{y,0}^{(\sigma_{1/2})} \quad \text{for the left bank}\,, \qquad n = 0\,,\\ &\frac{1}{2}\,(vd)_{N-1/2}\, c_{x,N} = \frac{1}{h}\,(Dd)_{N-1/2}\, c_{y,N}^{(\sigma_{N-1/2})} \quad \text{for the right bank}\,, \qquad n = N\,, \end{aligned} \tag{18}$$

where

$$c_{x,n} := (c_n^{k+1} - c_n^k)/\tau\,, \qquad k = 0, 1, \ldots, K-1\,,$$
$$c_n^k := c(x_k, y_n, D)\,, \qquad x_k := k\tau\,, \qquad y_n := nh\,, \qquad \tau := x_1/K\,, \qquad h := w/N\,,$$

i.e. τ and h are the discretization steps in direction of x and y, respectively, moreover,

$$c_{y,n} := (c_{n+1}^k - c_n^k)/h\,, \qquad c^{(\sigma)} := \sigma c^{k+1} + (1-\sigma)\, c^k\,,$$

where σ is a weighting parameter the choice of which constitutes the difference to the usual schemes.

The empirically given values of the functions $v(y)$, $d(y)$ were fitted by means of a standard routine for cubic spline interpolation [40]. By the resulting splines, equidistant values $v_{n+1/2}$, $d_{n+1/2}$ have been calculated for furcher use, according to the discretization scheme (17, 18).

In connection with (16) there are three numerical problems:

i) After appropriate reformulation, the background concentration and the point source define an initial distribution $c\,(0, y, D)$ which is discontinuous. To avoid physically senseless solutions, a monotone method is called for (i.e. a method converting monotone initial distributions into monotone final distributions).

ii) In accordance with the physical problem (Cl^- as well as Ca^{++} ions are so-called conservative substances, i.e. they are not degradable in the environment considered here), strict mass conservation must hold in the discrete model of the process.

iii) The coefficients in (16) degenerate at the boundary.

We remark firstly, concerning i), that the weighting parameter σ can be chosen uniquely (in dependence on τ, h, v, d and D) from the interval $[0, 1]$ so as to assure monotonicity of the difference scheme (17), (18) and closeness to the CRANK-NICOLSON case $\sigma = \frac{1}{2}$, see [26].

Next we consider in which kind mass conservation is guaranted by the scheme.

Summing up the equations (17) with respect to n and using (18), we get

$$0 = \sum_{n=1}^{N} (vd)_{n-1/2}\, \frac{h}{2}\, (c_{x,n} + c_{x,n-1}) = (Bc)_x\,, \tag{19}$$

where

$$Bc := \frac{h}{2}\left\{(vd)_{1/2}\, c_0 + \sum_{n=1}^{N-1} [(vd)_{n-1/2} + (vd)_{n+1/2}]\, c_n + (vd)_{N-1/2}\, c_N\right\}.$$

Hence, the discrete mass conservation law is

$$Bc = \text{const} = C_0\,,$$

that is Bc is, the quantity conserved exactly (up to rounding errors) by the discrete model (17), (18). Obviously, on smooth functions c,

$$Bc = \int_0^w vd\, \frac{\partial c}{\partial x}\, \mathrm{d}y + O(h^2)\,,$$

and here the integral is the quantity conserved in the continuous model (16).

Now, knowing that the input to the considered system consists in the background concentration β and the production of the point source of strength α, we have

$$C_0 = B\beta + \alpha = \beta\, \frac{h}{2}\left\{(vd)_{1/2} + \sum_{n=1}^{N-1} [(vd)_{n-1/2} + (vd)_{n+1/2}] + (vd)_{N-1/2}\right\} + \alpha\,. \tag{20}$$

To allow for several possibilities of application, we introduce a non-negative "shape" function δ of the discrete argument n with the property

$$\sum_{n=0}^{N} \delta(n) = 1 . \tag{21}$$

The cases of interest are

a) $\delta(n) = \delta_m(n) = \begin{cases} 0 & n \neq m \\ 1 & n = m \end{cases}.$

This models a small (point-wise) waste water tube ending in the interval $[y_{n-1/2}, y_{n+1/2}]$ for inner points $(1 \leqq m \leqq N-1)$, or in $[0, y_{1/2}]$ or $[y_{N-1/2}, 1]$ for the left and right banks, respectively.

b) $\delta(n) > 0 , \quad 0 \leqq n \leqq N' < N , \quad \delta(n) = 0 , \quad N' + 1 \leqq n \leqq N ,$

and this models either a great waste water inflow or a tributary, i.e. a source distributed over a finite interval. For instance, take

$$\delta(n) = \varkappa \frac{h}{2} \cdot \begin{cases} (vd)_{1/2} , & n = 0 \\ [(vd)_{n-1/2} + (vd)_{n+1/2}] , & 1 \leqq n \leqq N' - 1 \\ (vd)_{N'-1/2} , & n = N' \\ 0 , & N' + 1 \leqq n \leqq N \end{cases}$$

where

$$\varkappa = 1 \Big/ \sum_{1}^{N'} h(vd)_{n-1/2}$$

so as to satisfy (21). Now we may rewrite (20),

$$C_0 = B\beta + \sum_{n=0}^{N} \delta(n)\, \alpha = \frac{h}{2} \left\{ (vd)_{1/2} \left(\beta + \alpha\delta(0) \Big/ \frac{h}{2} (vd)_{1/2} \right) + \right.$$

$$\left. + \sum_{n=1}^{N-1} [(vd)_{n-1/2} + (vd)_{n+1/2}] \left(\beta + \alpha\delta(n) \Big/ \frac{h}{2} [(vd)_{n-1/2} + (vd)_{n+1/2}] \right) + (vd)_{N-1/2} \left(\beta + \alpha\delta(N) \Big/ \frac{h}{2} (vd)_{N-1/2} \right) \right\},$$

and have therefore, comparing with (19), the missing initial condition to (17), (18).

$$c_n^0 = \beta + \alpha\delta(n) \begin{cases} 1 \Big/ \frac{h}{2} (vd)_{1/2} , & n = 0 \\ 1 \Big/ \frac{h}{2} [(vd)_{n-1/2} + (vd)_{n+1/2}] , & 1 \leqq n \leqq N - 1 \\ 1 \Big/ \frac{h}{2} (vd)_{N-1/2} , & n = N . \end{cases} \tag{22}$$

This finishes the construction of the discrete cross-mixing model.

The tridiagonal system of linear equations defined by (17), (18), (22) is solved in the usual manner by shortened Gauss elimination, see, e.g., [19], p. 40. To cope with the steep gradients of (22), the above mentioned choice of σ is used which assures monotonicity and closeness to the second order scheme. Practically, it has proven to be sufficient to perform four steps in x-direction in this way. Then we switched to $\sigma = \frac{1}{2}$ economizing computational work. Alternatively but less accurate, $\sigma = 1$ could have been used to assure monotonicity and stability, see [1]. The resulting computing time for one calculation of $\{c(x_1, y_n, D)\}$ has been about 18 seconds (double precision, Algol programme, computer ROBOTRON EC 1040) for $N = 30$, $K = 34$.

From the above listened three numerical problems there remains to be discussed iii).

Parabolic equations with coefficients degenerating at the boundary arise, for instance, in heat conduction or diffusion problems posed in cylindrical or spherical coordinates. Second order difference schemes for such problems are adapted by using not only special coefficients but also special grids, see [19], p. 188 and p. 433, and [13]. In our context, due to the unknown function $D(y)$, it seems to be not reasonable to use special schemes. Rather, a general method such as (17), (18) is preferable. Therefore, the performance of (17), (18) was tested on smooth problems in cylindrical and spherical coordinates. In the tests the behaviour of the error was found to be of the type $O(h^t)$ with $t = 2.0$ for the error measured in the discrete L_2 norm, and with $t = 1.8$ for the maximum norm (for $20 \leqq N \leqq 160$, $K = N/2$, $\sigma = \frac{1}{2}$).

We mention that equation (16) may be considered to correspond to spherical coordinates if v and d have first order zeros at $y = w$ and $y = 0$, and if

$$D(y) = D_0(y)\, v(y) , \quad D_0(y) \geqq \text{const} > 0 . \tag{23}$$

Only this choice of $D(y)$ avoids (independently of the order of zeros of v and d) zeros or poles in the factor multiplying the second order derivative in (16) (after dividing this equation by vd), and leads to a first order singularity at $y = 0$ and $y = w$ in the factor multiplying the first derivative with respect to y.

5.3 Numerical solution of the inverse problem

The solution of the direct problem is now clear and we may turn to the inverse problem, to identify $D(y)$ from the given $\{\tilde{c}_j\}$. For this aim, $D(y)$ is represented in some n-parametric way with parameters $D_1, \dots, D_n$. These constant parameters are selected so as to have $D(y) > 0$ if $D_i > 0$, $i = 1, \dots, n$. The set of parameters $(D_1^*, \dots, D_n^*)$ minimizing the objective function $f_{III}(D_1, \dots, D_n)$,

$$f_{III}(D_1, \dots, D_n) = \sum_{j=1}^{m} (\tilde{c}_j - c(x_1, y_j, D_1 \dots D_n))^2 + M \sum_{i=1}^{n} (\max (0, \varepsilon - D_i))^2 \tag{24}$$

is considered to define the solution $D^*(y)$ of the inverse problem. As compared to f_{II} in (15), f_{III} contains a penalty term preventing negative D_i during minimization. For the constants M and ε in (24) we took 10 and 10^{-10}, respectively.

We are now in the standard situation of a nonlinear least squares problem. The specific algorithm used here for minimization of (24) has been developed from that given in [22]. It combines Newton and gradient steps to achieve effectiveness and robustness of the minimization. A broader description can be found in [24, 25].

5.4 Previous numerical experiments

The aim of [24, 25] was to study the influence of pseudorandom measurement errors on the identified coefficients. Since this has direct regard to the present work, we shall shortly restate here the main results (obtained by taking mean values over 5 to 20 experiments for fixed pairs (n, s) with n the number of parameters in (24) and s the standard deviation of the normally distributed pseudorandom errors), and add some evaluation and figures not reported in [24, 25]:

(a) Values of f_{III} at the minimum, f_{III}^*, divided by the number of freedoms $m - n$, turned out to have the magnitude of s^2, the variance of the pseudorandom errors. In other words, the residual variance of the model is of the order of the (simulated) measurement errors. This is illustrated by fig. 6, where the "noise amplification function"

$$\varphi = \sqrt{\frac{f_{III}^*}{(m-n)\, s^2}}$$

is seen to have two regions of different behaviour: The first one ($0 \leqq n/m < 0.15$, $\varphi < 1$) shows a decrease of φ with increasing s, which means that large measurement errors determine the model residual variance totally, while in case of small errors the model residual variance is dominated by the algorithmic errors. In the second region ($0.15 \leqq n/m \leqq 0.3$, $\varphi \approx 1$) the model residual variance is a sufficiently good representation of the measurement error, independent of its size. In a third region which is not shown in fig. 6, the values of φ decrease reaching zero (up to rounding errors) for $n = m$.

(b) There exists an optimal number of parameters, n, in the sense of a compromise between accuracy of identification and of stability against measurement errors. This number, n_{opt}, depends on the number of measurement points m, on the standard deviation s and also on the unknown coefficients (and is, hence, "unknownable" [2]). Maximum norm errors e_c of the identified coefficients are shown in fig. 7 for several n and s. All values of e_c, s, and f_{III}^* reported here in chapter 5.4 are relative ones.

(c) For $n = n_{opt}$, the maximum norm error of the identified coefficient was found to grow with s according to $e_c(s) = e_c(0) + 0.0531 \times s^{0.43}$, see fig. 8, where this least squares fit through the points $(s, e_c(s) - e_c(0))$ is plotted, $e_c(0) = 3.82 \cdot 10^{-4}$.

These are experimental results. But quite a number of theoretical papers points into the same direction, namely, that under certain conditions the heat conduction coefficient is determined by the values of $c(x_1, y)$ uniquely and in a stable manner, see e.g. [7, 8, 9, 15, 16], and the recent results of [31—33].

5.5 Numerical results

For the identification of $D(y)$, a polygonal representation of D has been tried, firstly. To use polygons is of no direct physical sense, but well suited to get an idea of the information content of the data set without imposing a modelling conception.

In the computations, the shape function introduced in 5.2 was chosen to be $\delta(n) = 0$ if $n = 1, \dots, N$ and $\delta(0) = 1$.

To get initial values of the parameters to be identified iteratively, we took firstly the ansatz $D(y) = D_1 = \text{const.}$ and $\beta = 270$ mg/l (compare with the concentration at the right hand bank in fig. 4). Furthermore we used the estimation

$$\alpha = Q_E \cdot \left(\frac{1}{m} \sum_{j=1}^{m} c_j - \beta \right) \approx 8 \text{ kg/s}$$

where $Q_E = 229.8$ m^3/s was the discharge of the Elbe river for the data set considered here.

The initial choice for D_1 was taken from the results of the Gaussian model (see chapter 4).

Starting with these data, by minimization of a functional f_{IV} similar to (24), the values were refined to

$$D_1^* = 0.0123 \text{ m}^2/\text{s}\,, \qquad \alpha^* = 8.61 \text{ kg/s}\,.$$

In all subsequent calculations α was fixed now on the above value which corresponds to an amount of 31 tons/hour (!).

Assuming a discharge Q_S of 50 m^3/s in the Saale river (this river represents the point source responsible for the transversal distribution of Cl^- ions under consideration), we obtain an excess Cl^- ions concentration of $\alpha/Q_S = 172$ mg/l in the Saale river, corresponding to an absolute concentration of $\beta + 172$ mg/l $= 442$ mg/l. (For comparison: oceanic waters exhibit Cl^- concentrations of about 19 g/l [38]).

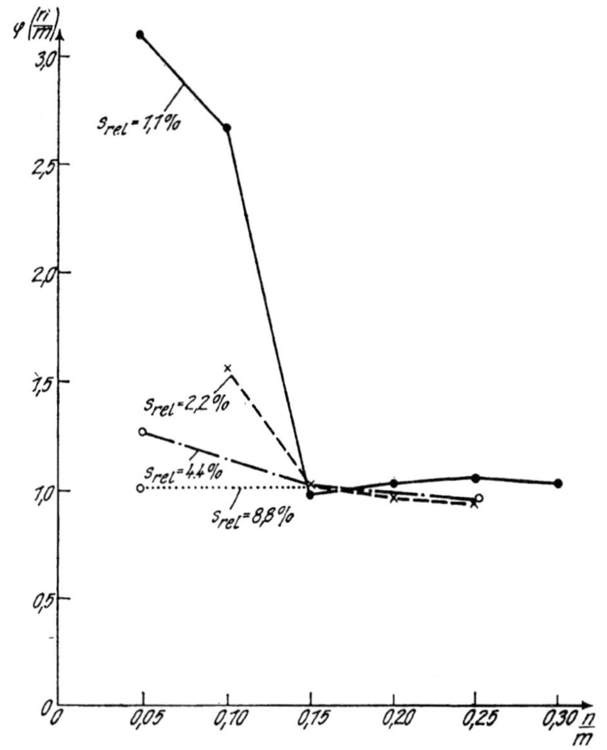

Fig. 6. Noise amplification function φ of the inverse problem m-number of measurements, n- number of free model parameters
The numbers 1.1%, 2.2% etc. denote values of the standard deviation s of the artificially generated measurement errors, relative to the mean value of the concentration.

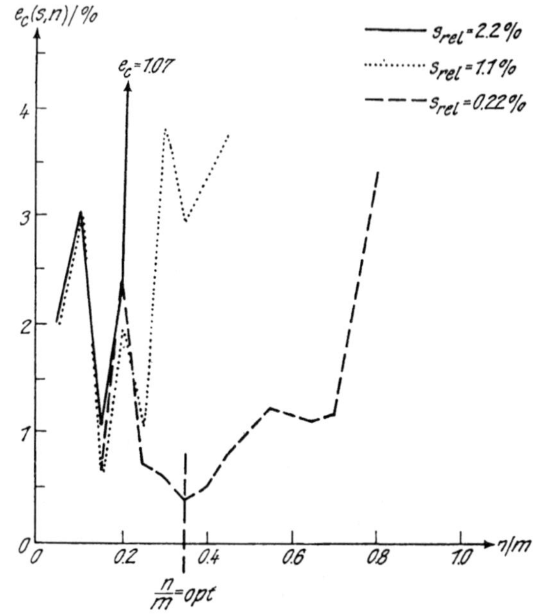

Fig. 7. Influence of the artificial measurement noise on the error of the parameters to be identified
s_{rel} — relative standard deviation of the artificially generated noise

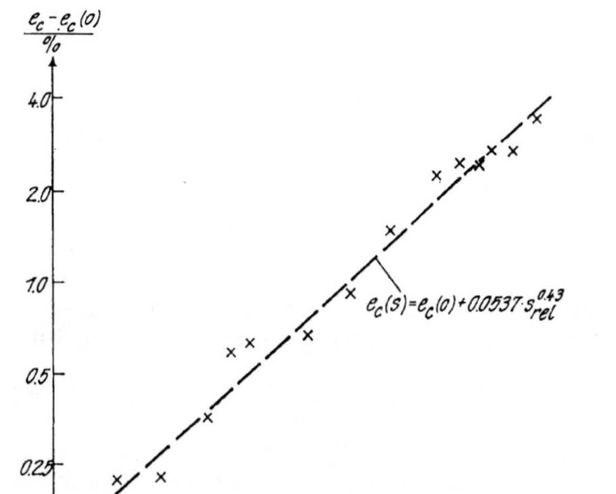

Fig. 8. Logarithmic presentation of the relative maximum norm error e_c of the identified coefficient, depending on the relative standard deviation s_{rel} of the artificially generated noise ——— Regression curve × × × Results of numerical experiments

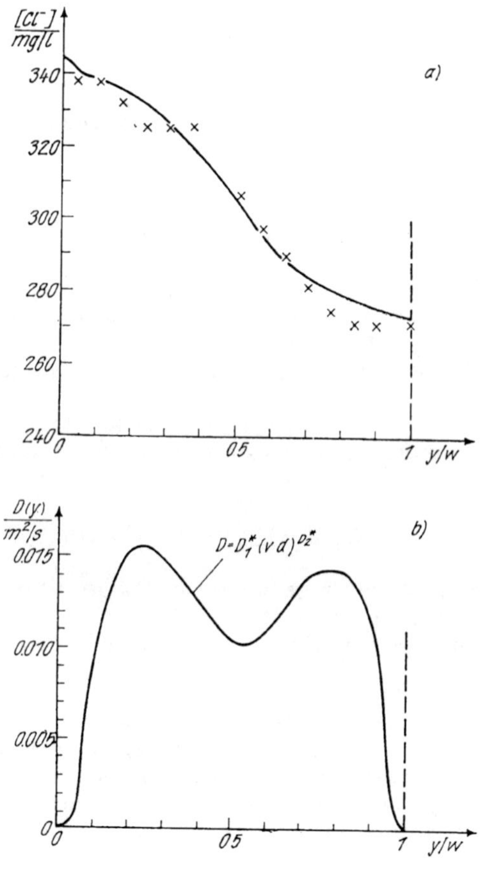

Fig. 9. Example of an identified mixing coefficient distribution (ansatz $D = D_1(v \cdot d)^{D_2}$) ▶
a) Calculated minimum variance concentration
b) Corresponding mixing coefficient distribution

Similarly as the point source strength α, the background concentration β might have been determined from the given data, too: The problem of finding an initial function for the heat conduction equation is improperly posed in general, but it is stable and with unique solution if some derivative of the initial function is known to be bounded [27, 28], which applies in our case since β is assumed to be constant.

Results for $D(y)$ represented by a polygon $P_n(y)$ with n knots in $(0, w)$, satisfying $P_n(0) = P_n(w) = 0$, are given in Table 1. Here

$$P_n(y) = \sum_{i=0}^{n+1} \max(D_i, \varepsilon_i) \cdot \psi\left((n+1)\cdot\frac{y}{w} - i\right), \qquad \psi(z) = \begin{cases} 0, & |z| > 1, \\ 1 - |z|, & |z| \leq 1, \end{cases}$$

$$\varepsilon_i = 10^{-10}, \qquad i = 1, \ldots, n, \qquad \varepsilon_0 = \varepsilon_{n+1} = D_0 = D_{n+1} = 0.$$

The final deviation from the measurement data is evaluated in percent according to the remark at the end of chapter 4. Table I in combination with fig. 7 shows that estimation with $n = 2$ and $n = 3$ is most preferable.

The results with piecewise linear D served as a first exploration of the measurement information. Subsequently, a number of other functions with several free parameters have been tried for representation of D.

Some computations were performed with polygons whose end points were not fixed to zero but free. The results (in the sense of the final values of f_{III}) were worse than in 5.4.3 for equal number of free parameters. Moreover, the small values of the gradient $\partial c/\partial y$ near the right hand bank led to badly determined values of $D(y)$ at $y = w$.

More successful turned out to be the choice (23) with $D_0(y) = D_1 =$ constant:

$$D = D_1 \cdot v(y), \qquad D_1^* = 0.0107.$$

Further computations have been carried out with functions containing two or three parameters and depending on v, d, on the solution c and the gradient $\partial c/\partial y$, see Table II. The nonlinear equations have been solved by the same difference scheme (17)—(18), combined with one or two iterations on the nonlinearity.

Table I. Results of polygonal approximations of the transverse mixing coefficients for the Cl^- data set
n — number of the unknown (free adjustable) parameters
m — number of the measured data (here: $m = 14$)
$m - n$ — degrees of freedom
D_{mv} — cross sectional mean value of $D(y)$

n	1	2	3	4	5	6
$\frac{\overline{\Delta c}}{\bar{c}}$	26%	12%	10%	10%	9%	8%
$\frac{\overline{\Delta c}}{\bar{c}}\sqrt{\frac{m-1}{m-n}}$	26,7%	13,3%	10,8%	12%	11,5%	11,1%
D_1^*	0.028	0.0220	0.0218	0.0200	0.0120	0.0080
D_2^*	—	0.0036	0.0074	0.0121	0.0256	0.0502
D_3^*	—	—	0.0045	0.0056	0.0060	0.0062
D_4^*	—	—	—	0.0046	0.0065	0.0085
D_5^*	—	—	—	—	0.0043	0.0036
D_6^*	—	—	—	—	—	0.0090
D_{mv}	0.0140	0.0085	0.0084	0.0085	0.0091	0.0122

Table II. Results of non-polygonial approximations of the transverse mixing coefficient for the Cl^- data set
n — number of unknown (free adjustable) parameters

n	$D = D(v(y), d(y))$	D_1^*	D_2^*	D_3^*	$\frac{\overline{\Delta c}}{\bar{c}}$
1	$D_1 =$ const.	0.0123	—	—	22%
1	$D_1 \cdot v \cdot d$	0.0107	—	—	15%
2	$D_1(v \cdot d)^{D_2}$	0.0081	2.391	—	14%
2	$(D_1 + D_2 c)\, v$	0.00816	0.134	—	9.8%
2	$\left(D_1 + D_2\left\lvert\frac{\partial c}{\partial y}\right\rvert\right) v$	0.00617	0.0272	—	14%
3	$(D_1 + D_2 c)\, v \Big/ \left(1 + D_3\left\lvert\frac{\partial c}{\partial y}\right\rvert\right)$	0.0128	0.2100	1.5	9.7%
3	$\left(D_1 + D_2\left\lvert\frac{\partial c}{\partial y}\right\rvert\right) v \Big/ (1 + D_3 c)$	0.00669	0.0161	$-1.57 \cdot 10^{-5}$	14%
2	$D_1 = \overline{D}$ Gaussian sol.	0.0097	—	—	15%

6. Comments and Conclusions

Table I gives the polygonal representation of the transverse mixing coefficient. The cross sectional mean values D_{mv} agree satisfactorily for all n. Particularly D_{mv} for $n = 6$ is in good accordance with the model $D_1 =$ const. (see the first line in Table II), but the mean square errors of both cases differ markedly (8% and 22%, respectively).

A comparison with fig. 7 suggests to consider the results to $n = 6$ as a case of "overestimation" where the gain in the functional value is due to better approximation of the noise, and it is surer to take $n = 2$ and $n = 3$. But obviously, the transverse non-homogeneities of $D(y)$ have a real (so difficultly to explain) physical meaning. The difference of the mean square errors of the model $D_1 = \text{const.}$ and the Gaussian solution (see Table II, 22% and 15%, respectively) is most probably founded by the fact, that the former is a one-parameter model while the Gaussian model has two parameters.

The model $D = D_1(v \cdot d)^{D_2}$ in Table II is motivated by (8) and physically reasonable. The ansatz $D = (D_1 + D_2 c)\, v$ can be considered as a variant of (23), whereas its physical sense is not quite clear, since turbulent diffusion is commonly scarcely influenced by Cl^- ions or other salt solutions. But it has been noted that very small concentrations of polymer additives can affect the eddy viscosity of flowing water [39], possibly responsible for the effect.

The model $D = (D_1 + D_2 c)\, v \Big/ \left(1 + D_3 \left|\frac{\partial c}{\partial y}\right|\right)$ is also not totally artificial. It can replace the mixing model with explicite time dependence obtained by fitting measurement data in [36] and avoids the breaking of temporal symmetry. On the other hand, time dependence of D can be shown to stem physically from the finitely small sizes and velocities of the eddies in turbulent flow (comp. [35]). To back nonlinear models like those considered in Table II, further measurement data should be analyzed. But we remark that several papers in [41] give reasons (referring back to [42]—[44]) for dependence of D on c or $\left|\frac{\partial c}{\partial y}\right|$. It can be noted also that the dependence of coefficients on the solution or its gradient is quite common in mechanics, see [36], [37] and further sources cited in [37]. (By the way, such problems with $1 + D_3 \left|\frac{\partial c}{\partial y}\right|$ in the denominator behave numerically friendly.)

From the physical point of view for us the model $D = D_1(v \cdot d)^{D_2}$ is the best understandable one, with good fitting properties, too. It can be used for predicting long-range and macro-scale effects of strong waste water loads in the Elbe river.

References

1 Андреев, В. Б., Об устойчивости по начальным данным разностных схем для параболических уравнений, Ж. вычисл. матем. и матем. физ. **11** (1971) 1462—1475.

2 Balakrishnan, A. V., Identification of distributed parameter systems: Non-computational aspects, in: Distributed Parameter Systems, Modelling and Identification (ed. A. Ruberti), Lecture Notes in Control and Information Sciences 1, Springer, Berlin 1978, 1—10.

3 Ball, J. A., Algorithms for RPN Calculators, Wiley-Interscience, New York 1978.

4 Baumert, H.; Peukert, V., Untersuchung der Quervermischung von Abwässern in Fließgewässern anhand von Sauerstoffprofilen, Wasserwirtschaft-Wassertechnik, Berlin, **25** (1975) 12, 38—45.

5 Baumert, H.; Becker, A., Water quality modelling in surface waters networks, IAHS-AISH Publ. No. 125 (1978) 269—276.

6 Baumert, H.; Heitmann, H.; Röbisch, D., Application of transverse mixing models to correct output data of automatic monitoring stations of river quality, Int. J. of Hydrol. Sci., Warsaw **6** (1979) No. 3—4, 161—169.

7 Безнощенко, Н. Я., Об определении коэффициента в параболическом уравнении, Дифф. Уравн. **X** (1974) № 1, 24—35.

8 Будак, Б. М.; Искендеров, А. Д., Об одном классе краевых задач с неизвестными коэффициентами, ДАН СССР **175** (1967) 1, 13—16.

9 Cannon, J. R., Determination of an unknown coefficient in parabolic differential equations, Duke Math. J. **30** (1963) 313—323.

10 Cannon, J. R., Determination of certain parameters in heat conduction problems, J. Math. Anal. Appl. 8 (1964) 188—201.

11 Cannon, J. R.,; Duchateau P., Determining unknown coefficients in an nonlinear heat conduction problem, SIAM J. Appl. Math. **24** (1973) 3, 298—314.

12 Dammann, J., Statistische Theorie der turbulenten Diffusion, Dtsch. Gewässerkdl. Mitt. **16** (1976) No. 3.

13 Фрязинов, И. В., О разностных схемах для уравнения Пуассона в полярной, цилиндрической и сферической системах координат, Ж. вычисл. матем. и матем. физ., **11** (1971) 5, 1215—1228.

14 Горошко, Б. И., Вычисление распределения концентрации в реке. Метеоролог. и Гидрол. (1972) № 9.

15 Jones, B. F., Determination of a coefficient in a parabolic differential equation, Part I: Existence and Uniqueness, J. Math. Mech. **11** (1962) 907—918.

16 Jones, B. F., Various methods for finding unknown coefficients in parabolic differential equations, Comm. Pure Appl. Math. **16** (1963) 33—44.

17 Монин, А. С.; Яглом, А. М., Статистическая Гидромеханика, Том I, II, Москва 1965, 1968.

18 Powell, M. J. D., A hybrid method for nonlinear equations, in: Numerical methods for nonlinear algebraic equations (ed. Ph. Rabinowich), Gordon and Breach 1970, 87—114.

19 Самарский, А. А., Теория разностных схем, Наука, Москва 1977.

20 Somlyódy, L., The use and limitation of a two-dimensional mixing model for rivers, IAHS-AISH Publ. No. 125 (1978) 96—105.

21 Sooky, A. A., Longitudinal dispersion in open channels, J. Hydraul. Div. Proc. ASCE **95** (1969) HY4.

22 Steene, N. M.; Byrne, G. D., The problem of minimizing nonlinear functionals, in: Least squares numerical solution of nonlinear algebraic equations (ed. G. D. Byrne & C. A. Hall), Academic Press, New York 1973, 185—240.

23 Stewart, G. W., A modification of Davidsons minimization method to accept finite difference approximations of derivatives, JACM **14** (1967) 72—83.

24 Stoyan, G., Numerical experiments on the identification of heat conduction coefficients, in: Theory of nonlinear operators (ed. R. Kluge), Akademie-Verlag, Berlin 1978, 259—268.

25 Stoyan, G., Identification of a spatially varying coefficient in a parabolic equation, a report on numerical experiments, in: Inverse and improperly posed problems for partial differential equations (ed. G. Anger), Akademie-Verlag, Berlin 1979, 249—258.

26 Stoyan, G., On a maximum-norm stable, monotone and conservative difference approximation of the one-dimensional diffusion-convection equation, in: Simulation der Migrationsprozesse im Boden- und Grundwasser (ed. L. Luckner), TU Dresden 1979, 139—160.

27 TYCHONOFF, A. N., Théorémes d'unicite pour l'équation de la chaleur, Mat. sb. **42** (1935) 2, 199–216.
28 TYCHONOFF, A. N., On the stability of inverse problems, Dokl. AN SSSR **39** (1943) 5, 176–179.
29 WARD, P. R. B., Transverse dispersion in oscillatory channel flow, J. Hydraul. Div. Proc. ASCE **100** (1974) HY6.
30 YOTSUKURA, N.; SAYRE, W. W., Transverse mixing in natural channels, Water Resources Research **12** (1976) No. 4.
31 KLUGE, R., An inverse problem for coefficients in linear equations. Uniqueness and iterative solution, in: Nonlinear Analysis, Theory and Applications, Proc. Summer School Berlin, DDR (ed.: R. KLUGE), Akademie-Verlag, Berlin 1981.
32 KLUGE, R., Zur starken Konvexität eines Funktionals bei der Koeffizientenbestimmung in linearen Operatorgleichungen, in: Proc. Autumn School "Variationsrechnung & optimale Steuerung", Vitte, 1980.
33 KLUGE, R., Zur Koeffizientenbestimmung in linearen partiellen Differentialgleichungen, Part I, Math. Nachrichten (in prep.).
34 GLOS, E.; FROTSCHER, D., BAUMERT, H., Analysis and simultaneous simulation of transient flow and matter transport scenarios for a lowlands river network using a generalized programme package, Proc. int. conf. "Numerical Modelling of River, Channel and Overland Flow" Bratislava, ČSSR, May 4–8, 1981.
35 KOLMOGOROV, A. N., A refinement of previous hypothesis concerning the local structure of turbulence in a viscous incompressible fluid at high Reynolds number, J. Fluid Mech. **13** (1962) 1, 82–85.
36 LANGENBACH, A., Variationsmethoden in der nichtlinearen Elastizitäts- und Plastizitätstheorie, Wiss. Ztschr. Humboldt-Univ. Berlin, Math.-Nat. R. IX (1959/1960) 145–164.
37 LANGENBACH, A., Monotone Potentialoperatoren in Theorie und Anwendung, VEB Dtsch. Verlag d. Wiss., Berlin 1976.
38 IVANOFF, A., Introduction a l'Oceanographie, Propriétés Physiques et Chimiques des Eaux de Mer, Tome I, Libraire Vuibert, Paris 1972.
39 SCHATZMANN, M. (ed.), Arbeits- und Ergebnisbericht des Sonderforschungsbereichs 80 für die Jahre 1977–1979, SFB 80/A/165, Universität Karlsruhe 1980.
40 ELLIS, T. M. R.; MCLAIN, D. H., Algorithm 514, A new method of cubic curve fitting using local data, ACM TOMS **3**, 2 (1977), 175–179.
41 HARRIS, C. J. (ed.), Mathematical modelling of turbulent diffusion in the environment, Acad, Press, 1979.
42 CORRSIN, S., Limitations of gradient transport models in random walk and in turbulence, Adv. Geophys. **18** A, 25.
43 ERDOGAN, M. E.; CHATWIN, P. C., The effect of curvature and buoyancy on the laminar dispersion of solute in a horizontal tube, J. Fluid Mech., **29** (1967), 465–484.
44 THATCHER, M. L.; HARLEMAN, D. R. F., A mathematical model for the prediction of unsteady salinity intrusion in estuaries, Rept. **144**, Dept. Civ. Eng. Mass. Inst. Techn. Cambridge, Massachusetts, USA.
45 CZERNUSZENKO, W., Unsteady diffusion of solutes in natural streams, Proc. int. conf. "Numerical Modelling of River, Channel and Overland Flow", Bratislava ČSSR, May 4-8, 1981.
46 CZERNUSZENKO, W., Masstransfer in the open chaunel flow, Journ of Hydrod. Sci. 2 (1975) No. 1–2, 19–35.
47 LANDAU L. D.; LIFSCHITZ E. W., Hydrodynamik, Akademie-Verlag, Berlin 1974.
48 BELTAOS, S., Transverse mixing tests in natural streams, J. Hydraulics Division ASCE 106 (1980) HY 10 1607–1625.

Engereicht am 13. April 1981

Anschriften: Dr. GISBERT STOYAN, Institut für Mathematik der Akademie der Wissenschaften der DDR, 1080 Berlin, Mohrenstr. 39 DDR
Dr. HELMUT BAUMERT, Institut für Wasserwirtschaft, DDR-1190 Berlin, Schnellerstraße 140

ZAMM 61, 629 – 631 (1981)

R. KÜHNAU

Verzögerung und virtuelle Masse bei Umströmung eines Hindernisses

Bringt man einen Körper in eine ebene Parallelströmung einer inkompressiblen reibungsfreien Flüssigkeit, wird die Strömung gestört, und die Zeit ändert sich, die ein Teilchen auf dem Wege zwischen zwei senkrecht zur ursprünglichen Strömungsrichtung liegenden Geraden benötigt. Es wird hier gezeigt, daß sich diese Zeit in einem gewissen Sinne vergrößert, wobei ein einfacher Zusammenhang mit der virtuellen Masse besteht.

If one puts a body into a plane parallel flow of an incompressible non-viscous liquid then the flow will be disturbed and the time which a particle needs between two lines orthogonal to the original direction of the flow will be changed. It will be shown here that this time is in some sense prolonged and that there is a simple connection to the virtual mass.

При погружении в плоский равномерный поток несжимаемой невязкой жидкости некоторого тела изменяются сам поток и время прохоэщения частицей пути между двумя ортогональными к направлению потока прямыми. Показано, что это время в определенном смысле увеличивается, причем устанавливается простая связь с виртуальной массой.

I. Wir betrachten reibungs- und wirbelfreie ebene Strömungen einer inkompressiblen Flüssigkeit ohne Zirkulation. Sei zunächst in einer vollständigen komplexen z-Ebene die Parallelströmung in Richtung der positiv reellen Achse mit konstantem Geschwindigkeitsvektor des Betrages 1 gegeben und die Zeit $R_2 - R_1$ ins Auge gefaßt, die ein Teilchen auf dem Wege von der Geraden $\Re z = R_1$ bis zur Geraden $\Re z = R_2 > R_1$ benötigt. Wie ändert sich diese Zeit, wenn — bei Beibehaltung dieser Geraden — ein Körper in diese Strömung eingebracht und dann stationär umströmt wird? Wir werden sehen, daß (in einem gewissen Sinne) sich diese Zeit i. a. vergrößert, und einen Zusammenhang mit konform-geometrischen Größen darstellen sowie mit der virtuellen Masse [1], [8], [9], [10], [11].

II. Zum Studium der hiermit zusammenhängenden Fragen sei nun in der z-Ebene ein von n Kontinuen berandetes Gebiet $\mathfrak{G}$ mit $z = \infty$ als innerem Punkt gegeben. $w(z)$ sei diejenige schlichte konforme Abbildung von $\mathfrak{G}$, die in $z = \infty$ durch

$$w(z) = z + \frac{a_1}{z} + \dots \tag{1}$$

hydrodynamisch normiert ist und bei der die Bildrandkomponenten Strecken parallel zur reellen Achse sind. Bekanntlich läßt sich $w(z)$ deuten als komplexes Potential zur stationären, reibungs- und wirbelfreien, inkompressiblen Durchströmung von $\mathfrak{G}$ ohne Zirkulation, wenn in $z = \infty$ der Geschwindigkeitsvektor (1, 0) ist. Das heißt: Die komplexe Zahl $\overline{w'(z)}$ repräsentiert jeweils den Geschwindigkeitsvektor. Sei $V = |w'(z)|$.

Die Gerade $\Re z = \text{const}$ wird wegen

$$w'(z) = 1 - \frac{a_1}{z^2} + \dots \tag{2}$$

bei hinreichend großem Wert für |const| von jeder Stromlinie genau einmal geschnitten. Fassen wir nun eine Stromlinie $\Im w = \text{const}$ zwischen $\Re z = -R$ und $\Re z = R$ bei derartig hinreichend großem R ins Auge, wobei wir solche Stromlinien ausschließen, die teilweise auf einer Randkomponente von $\mathfrak{G}$ verlaufen. Sei T die zur Durchlaufung benötigte Zeit. Dann wird bei Integration über dieses Stromlinienstück der Länge l bzw. das Bild (= Strecke) der Länge L in der w-Ebene:

$$T = \int \frac{|dz|}{V} = \int \left|\frac{dz}{dw}\right|^2 |dw| \,. \tag{3}$$

Die Schwarzsche Ungleichung bringt

$$T \geqq l^2/L \,. \tag{4}$$

Bei festem R strebt $T \to T_\infty = 2R$ für $|\Im w| \to \infty$ wegen (2) und (3). T_∞ ist auch die benötigte Zeit bei der ungestörten Strömung (ohne Anwesenheit der Randkomponenten von $\mathfrak{G}$). Nun ist

$$T - T_\infty \geqq \frac{l^2}{L} - T_\infty = \left(1 + \frac{l}{L}\right)(1 - T_\infty) - (L - T_\infty)\frac{l}{L}. \tag{5}$$

Hierbei strebt wegen (1) $L - T_\infty \to 0$ für $R \to \infty$. Ferner gilt dabei $l/L \to 1$: Es existiert nämlich zu gegebenem $\varepsilon > 0$ ein R_0 (dann festzuhalten), so daß für $|\Re w| > R_0$ stets $||dz/dw| - 1| < \varepsilon$, so daß diese Behauptung aus

$$l = \int |dz/dw|\, |dw| = \int_{-R_0}^{R_0} |dz/dw|\, |dw| + \int_{\mathfrak{z}} |dw| + \int_{\mathfrak{z}} (|dz/dw| - 1)\, |dw| \tag{6}$$

($\mathfrak{z}$ = Rest der Integrationsstrecke) folgt. Es besitzen $T - T_\infty$ und $1 - T_\infty \geqq 0$ wegen (3) und (6) offenbar für $R \to \infty$ einen Grenzwert, und es folgt damit:

Es gilt

$$\lim_{R\to\infty} (T - T_\infty) \geqq 2 \lim_{R\to\infty} (1 - T_\infty) \geqq 0 \tag{7}$$

mit Gleichheit genau für $w(z) \equiv z$, d. h. für den Fall, $\mathfrak{G}$ wird von Strecken parallel zur reellen Achse berandet.

Die Behauptung betreffs des Gleichheitszeichens folgt nicht etwa einfach durch Betrachtung von $1 - T_\infty$, da sehr wohl $1 - T_\infty$ nach 0 streben kann, ohne daß $w(z) \equiv z$. Man nehme z. B. ein von zwei bezüglich der reellen Achse zueinander spiegelbildlichen Randkomponenten berandetes Gebiet $\mathfrak{G}$, die reelle Achse dabei als Stromlinie. Die Diskussion des Gleichheitszeichens folgt aber durch feinere Betrachtung der Schwarzschen Ungleichung bei (4). In bekannter Weise folgt nämlich bei $|dz/dw| \not\equiv 1$ in Verschärfung von (4):

$$T \cdot L - l^2 \geqq p \cdot L \quad \text{mit} \quad p > 0\,, \qquad p \text{ unabhängig von } R\,.$$

Die linke Seite von (7) wird man in naheliegender Weise als „*Verzögerung*" der Strömung auf der betreffenden Stromlinie bezeichnen. In der Nähe einer umströmten Randkomponente kann man hierfür effektiv positive untere Schranken angeben, welche sich geometrisch und über bekannte allgemeine Verzerrungssätze bei konformen Abbildungen durch Abschätzung von $1 - T_\infty$ ergeben. Überdies liest man aus (3) ab, daß die Verzögerung nach $+\infty$ strebt, wenn die Stromlinie sich der umströmten Randkomponente nähert, falls einer der beiden Staupunkte an einem glatten Randstück auftritt.

Die Aussage, daß die Verzögerung $\geqq 0$ ist, gilt natürlich nicht bei endlichen Stücken von Feldlinien. Man betrachte z. B. eine Stromlinie in Umgebung einer Kante einer Randkomponente mit $V = \infty$.

III. Nach diesen nur auf eine Stromlinie bezogenen Überlegungen führen wir folgende mehr globale Betrachtung durch. Durch $-h \leqq \mathfrak{Im}\, w \leqq h$ (h hinreichend groß) grenzen wir eine Stromröhre um die umströmten Randkomponenten ab und von dieser Stromröhre durch $-R \leqq \mathfrak{Re}\, z \leqq R$ (R hinreichend groß) ein Teilstück. Wir betrachten für jedes Stromlinienstück $v = \mathfrak{Im}\, w = \text{const}$, $-R \leqq \mathfrak{Re}\, z \leqq R$, die zugehörige Zeit $T = T(v)$ und damit das Integral (vgl. (3))

$$F = \int_{-h}^{h} T(v)\, dv = \iint |dz/dw|^2\, du\, dv\,. \tag{8}$$

Dabei wird in der $u + iv = w$-Ebene über das entsprechende Bildflächenstück integriert (berandet außen von einer gewissen Kurve $\mathfrak{C}$, bestehend aus zwei zur reellen Achse parallelen Strecken $\mathfrak{s}_1$ und $\mathfrak{s}_2$ und zwei die Endpunkte verbindenden Kurvenstücken, und innen von n zur reellen Achse parallelen Schlitzen). F stellt also den in der Strömungsebene abgegrenzten Flächeninhalt außerhalb der Randkomponenten, die insgesamt den (äußeren) Inhalt I ausmachen mögen, dar und erfüllt demnach auch

$$F + I = -\frac{1}{2i}\int_{\mathfrak{c}} z\, d\bar{z} = -\frac{1}{2i}\int_{\mathfrak{C}} z(w)\, \overline{(dz/dw)}\, d\bar{w}\,. \tag{9}$$

Dabei sei $\mathfrak{c}$ das Urbild von $\mathfrak{C}$ in der z-Ebene, wie $\mathfrak{C}$ positiv orientiert.

Wir entwickeln nun um den unendlich fernen Punkt

$$w(z) = z + \frac{a_1}{z} + \frac{a_2}{z^2} + \dots\,, \qquad z(w) = w + \frac{A_1}{w} + \frac{A_2}{w^2} + \dots\,, \qquad A_1 = -a_1\,. \tag{10}$$

Nach dem Bieberbach-Gronwallschen Flächensatz ist

$$|A_n| \leqq \varrho^{n+1}/\sqrt{n}\,, \tag{11}$$

wenn $|w| = \varrho$ alle Randkomponenten in der w-Ebene umschlingt. Damit wird

$$\begin{aligned} F + I &= -\frac{1}{2i}\int_{\mathfrak{C}} \left(w - \frac{a_1}{w} + \frac{A_2}{w^2} + \frac{A_3}{w^3} + \dots\right)\left(1 + \frac{\bar{a}_1}{\bar{w}^2} - \frac{2\bar{A}_2}{\bar{w}^3} + \dots\right) d\bar{w}\,, \\ &= -\frac{1}{2i}\int_{\mathfrak{C}} w\, d\bar{w} + \frac{a_1}{2i}\int_{\mathfrak{C}} \frac{d\bar{w}}{w} - \frac{\bar{a}_1}{2i}\int_{\mathfrak{C}} \frac{w}{\bar{w}^2}\, d\bar{w} + N\,, \end{aligned} \tag{12}$$

mit

$$|N| \leqq \sum_{n=2}^{\infty} n\,|A_n| \int_{\mathfrak{C}} \frac{|dw|}{|w|^n} + \frac{1}{2}\sum_{k,\,l=1}^{\infty} l\,|A_k|\,|A_l| \int_{\mathfrak{C}} \frac{|dw|}{|w|^{k+l+1}}\,.$$

Nach (11) folgt demnach

$$|N| \leqq C/h^2 \quad \text{mit von } R \text{ und } h \text{ unabhängigem } C.$$

Man errechnet noch, da auf $\mathfrak{s}_1$ und $\mathfrak{s}_2$ gilt $d\bar{w} = dw$ und $\bar{w} = w - 2ih$ bzw. $= w + 2ih$:

$$\int_{\mathfrak{C}} \frac{d\bar{w}}{w} = 2\pi i + P_1\,, \qquad \int_{\mathfrak{C}} \frac{w\, d\bar{w}}{\bar{w}^2} = \int_{\mathfrak{s}_1} \frac{w\, dw}{(w - 2ih)^2} + \int_{\mathfrak{s}_2} \frac{w\, dw}{(w + 2ih)^2} + P_2 = -2\pi i + P_3$$

mit $|P_k| \leqq ch/R$, wobei c von R und h unabhängig ist. Damit wird nach (8) und (12)

$$\int_{-h}^{+h} [T(v) - T_\infty]\, dv = 2\pi\, \mathfrak{Re}\, a_1 - I + N + P_4\,.$$

Wir fassen nun in naheliegender Weise

$$G = \lim_{h\to\infty} \left(\lim_{R\to\infty} \int_{-h}^{+h} [T(v) - T_\infty]\, dv \right) \tag{13}$$

auf als *Gesamtverzögerung G, die die Strömung erfährt.* Dann ergibt unsere Rechnung

$$G = 2\pi\, \Re e\, a_1 - I\,. \tag{14}$$

Den gleichen Wert erhält man bekanntlich für die u. a. in [1], [8], [9], [10], [11] (auch im räumlichen Analogon) betrachtete virtuelle Masse.

Wir haben damit den

Satz. *Die Gesamtverzögerung stimmt mit der virtuellen Masse überein.*

Wir schließen noch einige einfache Bemerkungen an, die die im wesentlichen bekannten Eigenschaften der Gesamtverzögerung G bzw. der virtuellen Masse in Zusammenhang mit der Theorie der konformen Abbildung bringen.

IV. Wegen (14) gilt nach [3] (Ungleichung (5) auf S. 31), [12] (S. 397) $G \geqq 0$ mit Gleichheit genau dann, wenn die Berandung von $\mathfrak{G}$ aus zur reellen Achse parallelen Strecken besteht, also die ursprüngliche Parallelströmung nicht gestört wird. Im Fall der virtuellen Masse ist diese Aussage natürlich evident.

V. Nach H. GRÖTZSCH (vgl. z. B. [4], Theorem 6.10) ergibt sich sofort folgendes. Der Koeffizient von $1/z$ bei *allen* hydrodynamisch normierten schlichten konformen Abbildungen von $\mathfrak{G}$ variiert in einer gewissen Kreisscheibe mit Mittelpunkt m und Durchmesser $2r$ (nach M. SCHIFFER „Spanne" genannt). Entsteht $\mathfrak{G}_\alpha$ aus $\mathfrak{G}$ durch Drehung um den Winkel α (um irgendeinen Drehpunkt) und wird $a_{1\alpha}$ der Entwicklungskoeffizient bei $\mathfrak{G}_\alpha$, wie a_1 bei $\mathfrak{G}$, so gilt:

$$a_{1\alpha} = m\, e^{2i\alpha} + r\,. \tag{15}$$

Setzt man also $m = |m|\, e^{i\beta}$, so wird für die Gesamtverzögerung G_α bei $\mathfrak{G}_\alpha$:

$$G_\alpha = 2\pi r - I + 2\pi\, |m| \cos(\beta + 2\alpha)\,. \tag{16}$$

Dies entspricht dem bekannten Tensorcharakter der virtuellen Masse. Insbesondere folgt aus (16) nach Mittelung über α, daß die mittlere Gesamtverzögerung (= mittlere virtuelle Masse) gleich der mit π multiplizierten Spanne ist, vermindert um den Flächeninhalt I.

VI. Nach einem Gedanken von B. EPSTEIN (vgl. z. B. die Literatur in [6] und dort Satz 1 bzw. Ungleichung (12)), angewandt auf $z(w)$, ergibt sich

$$I \leqq \pi r - \pi r^{-1} |a_1 - r|^2\,,$$

also

$$I(1 - \sqrt{\ })/(1 + \sqrt{\ }) \leqq G \leqq I(1 + \sqrt{\ })/(1 - \sqrt{\ }) \quad \text{mit} \quad \sqrt{\ } = \sqrt{1 - \frac{I}{\pi r}}\,. \tag{17}$$

Das Gleichheitszeichen steht z. B. im Falle nur einer Randkomponente genau für jeweils achsenparallel liegende Ellipsen, wobei sich das Achsenverhältnis eindeutig aus I/r bestimmt. Allein bei Kenntnis von I ist bekanntlich keine Abschätzung für G möglich, wie das Beispiel von Ellipsen zeigt.

Weitere Abschätzungen für G bzw. die virtuelle Masse ergeben sich aus [9], [10], [11] sowie nach [5] (Satz 2 mit $\nu \equiv 0$) und [7] (rechte Seite der Ungleichung (24) mit $Q \to \infty$), indem dort irgendein zulässiges μ eingesetzt wird.

VII. Man zeichne in der z-Ebene eine etwa analytische geschlossene Jordankurve C, die den Inhalt I umschlingt. Die Gesamtverzögerung bei Umströmung des Äußeren von C ist $2\pi\, \Re e\, a_1 - I$. Wird jetzt an irgendeiner Stelle von C ein Stückchen herausgebrochen, so entsteht aus dem Äußeren $\mathfrak{G}$ von C ein Schlitzgebiet $\mathfrak{G}^*$ (und das Innere von C wird mit durchströmt). Die hierzu gehörige Gesamtverzögerung konvergiert, wenn die Länge des herausgebrochenen Stückchens nach Null strebt, nach allgemeinen Konvergenzsätzen der Theorie der konformen Abbildungen gegen $2\pi\, \Re e\, a_1$ (vgl. [2], S. 195 ff.). Durch derartige Entfernung eines „infinitesimalen Stückchens" aus C erleidet also die Gesamtverzögerung bzw. die virtuelle Masse eine sprungartige Erhöhung um I.

Literatur

1 BERGMAN, S.; SCHIFFER, M., Kernel functions and elliptic differential equations in mathematical physics, New York 1953.
2 GOLUSIN, G. M., Geometrische Funktionentheorie, Berlin 1957.
3 GRUNSKY, H., Koeffizientenbedingungen für schlicht abbildende meromorphe Funktionen, Math. Z. **45** (1939), S. 29–61.
4 JENKINS, J. A., Univalent functions and conformal mapping, Berlin-Göttingen-Heidelberg 1958.
5 KÜHNAU, R., Gauss-Thomsonsches Prinzip minimaler Energie, verallgemeinerte transfinite Durchmesser und quasikonforme Abbildungen, Lect. Notes Math. **743**, Berlin/Heidelberg/New York 1979, S. 140–164.
6 KÜHNAU, R., Die Spanne von Gebieten bei quasikonformen Abbildungen, Arch. Rat. Mech. Anal. **65** (1977), S. 299–303.
7 KÜHNAU, R., Verzerrungsaussagen bei quasikonformen Abbildungen mit ortsabhängiger Dilatationsbeschränkung und ein Extremalprinzip der Elektrostatik in inhomogenen Medien, Comm. Math. Helv. **53** (1978), S. 408–428.
8 PÓLYA, G.; SZEGÖ, G., Isoperimetric inequalities in mathematical physics, Princeton 1951.
9 SCHIFFER, M., Sur la polarisation et la masse virtuelle, Comptes rendus Paris **244** (1957), S. 3118–3121.
10 SCHIFFER, M., Problèmes aux limites et fonctions propres de l'équation intégrale de Poincaré et Fredholm, Comptes rendus Paris **245** (1957), S. 18–21.
11 SCHIFFER, M.; SZEGÖ, G., Virtual mass and polarization, Trans. Amer. Math. Soc. **67** (1949), S. 130–205.
12 TSUJI, M., Potential theory in modern function theory, Tokyo 1959.

Eingereicht am 27. Februar 1981

Anschrift: Dozent Dr. REINER KÜHNAU, Sektion Mathematik der Martin-Luther-Univ. Halle-Wittenberg, 4020 Halle an der Saale, Universitätsplatz 6, DDR

ZAMM 61, 633 – 638 (1981)

F. WEIDENHAMMER

Nicht-lineare Schwingungen mit fast-periodischer Parametererregung

Herrn Professor Dr. S. FALK zur Vollendung seines 60. Lebensjahres gewidmet.

Bekanntlich kann auch bei fast-periodischer Parametererregung die Ruhelage für gewisse Frequenzen instabil werden. Die ersatzweise eintretenden nicht-linearen Schwingungen werden näherungsweise für einige dieser Frequenzen berechnet.

It is well-known that even in the case of almost periodic parametrical excitations the zero solution might become unstable for certain frequencies initiating non-linear vibrations which are calculated approximately for some of these frequencies.

Как известно, положение покоя может быть неустойчивым и в случая почти-периодического параметрического возбуждения при определенных частотах. Возникающие нелинейные колебания вычисляются приближенно для некоторых из этих частот.

1. Fragestellung

Es ist bekannt, daß lineare Schwinger mit periodischer [1], fast-periodischer [2], [7] oder stochastischer [3] Parametererregung Instabilitäten der Ruhelage zeigen. Die ersatzweise dann eintretenden, stationären Schwingungen lassen sich nur nicht-linear berechnen und bieten daher selbst wieder Stabilitätsprobleme. Diese nicht-linearen parametererregten Schwingungen sollen für einige wichtige Instabilitäten der Ruhelage bei fastperiodischer Erregung nachfolgend mit den ersten Schritten einer Störungsrechnung genähert angegeben werden.

2. Ausgangsgleichungen

Die fast-periodische Parametererregung in

$$q^2y'' + y + 2Dq\varepsilon^n y' + \varepsilon\beta_1 \cos\tau y + \varepsilon \sum_{j=2}^{N} \beta_j \cos(\alpha_j\tau - \psi_j)\, y + \gamma\varepsilon^n y^3 = 0 \tag{1}$$

sei speziell mit den N, endlich vielen, inkommensurablen, dimensionslosen Frequenzen α_j gegeben, und sie wird dann auch quasiperiodisch genannt. In (1) bezeichnen Striche Ableitungen nach der dimensionslosen Zeitzählung

$$\tau = \omega_1 t\,,$$

$D < 1$ das Lehrsche Dämpfungsmaß, β_j und ψ_j die Amplituden und Phasen der Parametererregung und $\gamma > 0$ die kubische Überlinearität. Mit $\varepsilon \ll 1$ bei $n = 1, 2$ wurde ein Kleinheitsparameter eingeführt, und mit der Eigenfrequenz ω_e von (1) wird das Frequenzverhältnis

$$q = \frac{\omega_1}{\omega_e} = \frac{\omega}{\omega_e} \tag{2}$$

in seiner Abhängigkeit von den Daten der Parametererregung betrachtet. In (2) wird zum Beispiel mit $\omega_1 := \omega$ die tiefste Parameterfrequenz gegenüber den anderen, inkommensurablen $\omega_j = \alpha_j\omega$ ($j = 2, \dots, N$) bevorzugt und einfach mit ω bezeichnet. Dabei wird vorausgesetzt, daß alle diese Frequenzen einander proportional sind. So wird dies zum Beispiel der Fall sein, wenn diese Frequenzen von den Eigenfrequenzen eines Kontinuums herrühren. Im übrigen kann man sich in (2) durch bloße Umnumerierung auf jedes ω_j beziehen. Mit dem in der Theorie der Parametererregung vielfach üblichen Frequenzverhältnis (2) soll der Vergleich mit den Ergebnissen der streng periodischen Parametererregung ermöglicht werden, wobei sich stets die umfassende Bedeutung der rein periodischen Erregung zeigen wird.

3. Nicht-lineare Schwingungen

Von den durch (1) beschriebenen Schwingern weiß man, in der Nähe welcher Parametererregungsfrequenzen die Ruhelage instabil wird [4]. Der wichtigste, weil breiteste Frequenzbereich liegt bei $\omega = 2\omega_e$ und seine Breite ist der ersten Potenz von ε proportional. Die dann ersatzweise eintretende, nicht-lineare Schwingung kann daher als eine fast-periodische Lösung von (1) mit $n = 1$ über eine Störungsrechnung gefunden werden. Entwickelt man zu dem Zweck y und ω nach Potenzen von ε und bezeichnet die Anteile von y mit y_i, die in (2) von ω auf ω_e bezogenen Anteile mit q_i, so gelten allgemein die Entwicklungen

$$y = y_0 + y_1\varepsilon + y_2\varepsilon^2 + \dots\,, \qquad q = \frac{\omega}{\omega_e} = q_0 + q_1\varepsilon + q_2\varepsilon^2 + \dots\,. \tag{3), (4}$$

Doppelte Eigenfrequenz in der Nähe einer Parameterfrequenz.

Für den genannten wichtigsten Bereich bei $2\omega_e = \omega$ ist also $q_0 = 2$, und man hat mit (3) und (4) für die Potenzen ε^0 und ε^1 die Gleichungen

$$4y_0'' + y_0 = 0\,, \tag{5,0}$$

$$4y_1'' + y_1 = -4q_1y_0'' - 4Dy_0' - \beta_1 \cos\tau y_0 - \sum_{j=2}^{N} \beta_j \cos(\alpha_j\tau - \psi_j)\, y_0 - \gamma y_0^3\,, \dots \tag{5,1}$$

rekursibel zu lösen. Aus (5,0) erhält man dafür mit den Integrationskonstanten A_0 und ϑ_0 zunächst

$$y_0 = A_0 \cos\left(\frac{\tau}{2} - \vartheta_0\right) \tag{6}$$

und diese Lösung führt in (5,1) rechtsseitig auf Terme $\cos(\tau/2 - \vartheta_0)$ und $\sin(\tau/2 - \vartheta_0)$, deren Vorfaktoren also zur Vermeidung säkularer Lösungsterme in y_1 zu Null abgeglichen werden müssen. Sie verlangen daher für A_0, ϑ_0:

$$q_1 A_0 - \frac{\beta_1}{2} A_0 \cos 2\vartheta_0 - \frac{3}{4}\gamma A_0^3 = 0 \tag{7}$$

und

$$2DA_0 + \frac{\beta_1}{2} A_0 \sin 2\vartheta_0 = 0\,. \tag{8}$$

Nach Lösung von (7) und (8) kann dann y_1 aus

$$4y_1'' + y_1 = -\sum_{j=2}^{N} \frac{\beta_j}{2} A_0 \left\{\cos\left[\left(\alpha_j + \frac{1}{2}\right)\tau - \psi_j - \vartheta_0\right] + \cos\left[\left(\alpha_j - \frac{1}{2}\right)\tau - \psi_j + \vartheta_0\right]\right\} - \\ - \left(\frac{\beta_1}{2} A_0 + \frac{1}{4}\gamma A_0^3\right)\cos\left(\frac{3}{2}\tau - 3\vartheta_0\right) \tag{9}$$

problemlos berechnet werden. Zu der inhomogenen Lösung von (9) muß sodann die homogene Lösung entsprechend (6) nun mit freien A_1, ϑ_1 zugefügt werden, so daß das Verfahren mit neuerlichem Säkulargliedabgleich fortführbar ist. Doch hat wohl nur die aus(7) und (8) berechenbare erste Näherung praktische Bedeutung, und Konvergenzfragen bleiben offen. Man erkennt in (8) und (9) zunächst wieder in $A_0 = 0$ die Ruhelage als eine mögliche Lösung. Da diese instabil werden kann, muß es ersatzweise mindestens eine stabile Schwingung geben. Abhängig von den Näherungen für die Amplitude A_0 und Phase ϑ_0 läßt sich die zugehörige Parameterfrequenz ω über q_1 aus (4) nicht-linear amplitudenabhängig berechnen, wie dies für nicht-lineare Schwingungen typisch ist. Zu dem Zweck erhält man zunächst durch Elimination der Phase ϑ_0 aus (7) und (8)

$$q_1 = \frac{3}{4}\gamma A_0^2 \pm \sqrt{\frac{\beta_1^2}{4} - 4D^2}$$

und damit ergeben sich in der ersten Näherung auch in vorliegendem Falle zunächst wieder zwei Frequenzgleichungen

$$q = \frac{\omega}{\omega_e} = 2 + \frac{3}{4}\gamma\varepsilon A_0^2 \pm \varepsilon\sqrt{\frac{\beta_1^2}{4} - 4D^2} + \dots, \tag{10}$$

wie aus der Theorie der harmonischen Parametererregung gut bekannt ist.

Im ungedämpften Fall ist $D = 0$ und man kann aus (8) dann nur $\vartheta_0 = 0$ und $\vartheta_0 = \pi/2$ als wesentlich entnehmen. Nach (7) gehört daher zur Grundlösung (6) $y_0 = A_0 \cos\tau/2$ das obere Wurzelvorzeichen in (10), während mit $\vartheta_0 = \pi/2$ die andere Grundlösung $y_0 = A_0 \sin\tau/2$ zum unteren Vorzeichen gehört. In der Näherung (10) kommt bemerkenswerter Weise der Einfluß der Fastperiodizität noch nicht vor. Somit lassen sich in dieser Näherung auch die Stabilitätsentscheidungen mit der Theorie der Mathieu-Gleichung allein erledigen, und alle Ergebnisse können der Literatur entnommen werden. Man findet so auch das bekannte Ergebnis wieder, daß ersatzweise für die instabile Ruhelage nur eine nicht-lineare parametererregte Schwingung stabil ist. So ergibt im vorliegenden Falle eine Störungsrechnung für $y = y_0 + \eta$ in (1), daß die ersten Potenzen der Störung η nur für den Typ $y_0 = A_0 \sin\tau/2$ beschränkt bleiben, und daher haben nur dieser Typ und das zugehörige untere Vorzeichen in (10) physikalische Bedeutung (Abb. 1). Die elementare Rechnung hierzu soll hier nicht wiedergegeben werden, doch hat man beim Vergleich mit der Literatur auf den Phasenansatz der Parametererregung (hier: $+\varepsilon\cos\tau$) und auf den Charakter der Nichtlinearität (hier über-linear: $\gamma > 0$) zu achten.

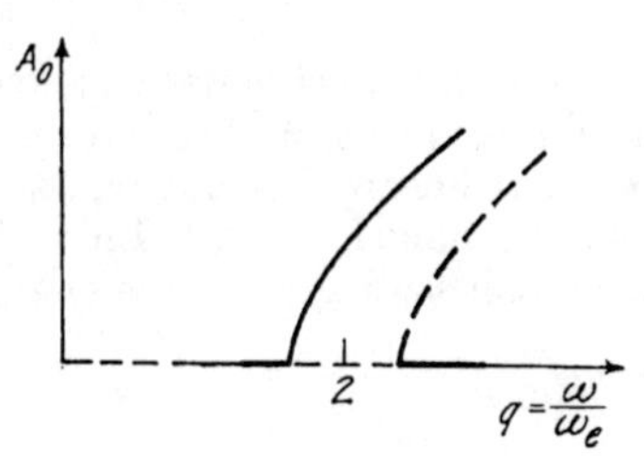

Abb. 1 Amplitude A_0 abhängig von Parameterfrequenz ω, wenn Ruhelage bei $2\omega_e \approx \omega$ instabil

Eigenfrequenz in der Nähe einer Parameterfrequenz

Im Instabilitätsbereich $\omega_e = \omega$ sind bekanntlich alle Effekte nur von der Größenordnung ε^2, und es gehen nun sämtliche Amplituden β_j der Parametererregung von (1) schon in die Berechnung der ersten Näherung ein [4].

Die ersatzweise eintretende, nicht-lineare Schwingung enthält allerdings die Nichtlinearität γ schon in der ersten Potenz ε, wie man von entsprechender periodischer Parametererregung her weiß [5]. Da dieser Effekt bekannt ist, wird hier zur Vereinfachung der formalen Rechnung die Nichtlinearität mit ε^2 klein angesetzt ($n = 2$ in (1)), was aber beim Vergleich mit der Literatur leicht berücksichtigt werden kann und phänomenologisch keine

Einschränkung bedeutet. Die Dämpfung darf allerdings in (1) zwingend nur von der Größenordnung ε^2 klein sein, da anderenfalls jegliche Instabilität der Ruhelage unterdrückt wird. Mit diesen Voraussetzungen über die Größenordnungen folgen hier aus (1) bis (4) die Gleichungen

$$q_0^2 y_0'' + y_0 = 0\,, \tag{11,0}$$

$$q_0^2 y_1'' + y_1 = -\beta_1 \cos\tau y_0 - \sum_{j=2}^{N} \beta_j \cos(\alpha_j \tau - \psi_j)\, y_0\,, \tag{11,1}$$

$$q_0^2 y_2'' + y_2 = -2q_0 q_2 y_0'' - 2Dq_0 y_0' - \beta_1 \cos\tau y_1 - \sum_{j=2}^{N} \beta_j \cos(\alpha_j\tau - \psi_j)\, y_1 - \gamma y_0^3\,. \tag{11,2}$$

Außer der Ruhelage ($A_i = 0$, $i = 0, 1, \dots$) haben sie für hier $q_0 = 1$ die Schwingungslösungen mit den Integrationskonstanten A_0, ϑ_0 und A_1, ϑ_1:

$$y_0 = A_0 \cos(\tau - \vartheta_0)\,, \tag{12,0}$$

$$\begin{aligned} y_1 = {} & A_1 \cos(\tau - \vartheta_1) - \frac{\beta_1}{2} A_0 \cos\vartheta_0 + \frac{\beta_1}{6} A_0 \cos(2\tau - \vartheta_0) - \\ & - \sum_{j=2}^{N} \frac{\beta_j}{2} A_0 \frac{\cos[(\alpha_j - 1)\tau - \psi_j + \vartheta_0]}{-(\alpha_j - 1)^2 + 1} - \sum_{j=2}^{N} \frac{\beta_j^2}{2} A_0 \frac{\cos[(\alpha_j + 1)\tau - \psi_j - \vartheta_0]}{-(\alpha_j + 1)^2 + 1}\,. \end{aligned} \tag{12,1}$$

In (11,2) verlangen nun y_0 und y_1 rechtsseitig zum Abgleich der Säkularglieder für A_0 und ϑ_0 die Gleichungen:

$$2q_2 + \frac{\beta_1^2}{6} + \frac{\beta_1^2}{4} \cos 2\vartheta_0 + \sum_{j=2}^{N} \frac{\beta_j^2}{2} \frac{1}{4 - \alpha_j^2} - \frac{3}{4} \gamma A_0^2 = 0\,, \tag{13}$$

$$2D - \frac{\beta_1^2}{4} \sin 2\vartheta_0 = 0\,. \tag{14}$$

Durch Elimination der Phase ϑ_0 erhält man aus ihnen:

$$q = \frac{\omega}{\omega_e} \doteq 1 + \left(-\frac{\beta_1^2}{12} \mp \frac{1}{2} \sqrt{\frac{\beta_1^2}{16} - 4D^2} - \sum_{j=2}^{N} \frac{\beta_j^2}{4} \frac{1}{4 - \alpha_j^2} + \frac{3}{8} \gamma A_0^2 \right) \varepsilon^2 + \dots\,. \tag{15}$$

Dazu entnimmt man aus (14) für ungedämpfte Schwingungen ($D = 0$), daß (12,0) mit $\vartheta_0 = 0$ entweder einen Kosinus-Typ oder mit $\vartheta_0 = \pi/2$ einen Sinus-Typ beschreibt, und ihnen hat man in der Frequenz-Amplituden-Relation (15) jeweils das obere ($\vartheta_0 = 0$) oder das untere ($\vartheta_0 = \pi/2$) Vorzeichen zuzuordnen. Alle Terme von (15) sind je für sich bekannt. So hat man für $\beta_j = 0$, $j = 2, \dots, N$ streng harmonische Parametererregung ([5] mit Nichtlinearität proportional ε^2) und für den linearen Schwinger mit fast-periodischer Parametererregung hat man $\gamma = 0$ zu setzen [4]. Da die Ruhelage instabil werden kann, wird man eine der beiden durch (15) und (12,0) gegebenen Schwingungen als ersatzweise stabil erwarten. Um diese Stabilitätsentscheidung zu leisten, geht man mit dem Störungsansatz $y_0 + \eta$ in (1) ein und erhält nach Linearisierung bezüglich der Störung η zur Prüfung der Stabilität der Näherung y_0

$$q^2\eta'' + \eta + 2Dq\varepsilon^2\eta' + \varepsilon\beta_1 \cos\tau\eta + \varepsilon \sum_{j=2}^{N} \beta_j \cos(\alpha_j\tau - \psi_j)\,\eta + 3\gamma\varepsilon^2 y_0^2 \eta = 0\,. \tag{16}$$

In dem Lösungsansatz

$$\eta(\tau) = e^{i\lambda\tau}\Phi(\tau) \tag{17}$$

werde in Anlehnung an die Theorie von G. FLOQUET λ der charakteristische Exponent genannt und $\Phi(\tau)$ sei eine fast-periodische Funktion. Somit entscheidet dann λ über das jeweilige Wachstumsverhalten und damit über die Stabilität, während die in $\Phi(\tau)$ eingebrachten Integrationskonstanten die Allgemeinheit der Störung sichern. Zur Berechnung von λ bei $q_0 = 1$ werden wieder die Entwicklungen

$$\lambda = \lambda_2\varepsilon^2 + \dots\,, \qquad \Phi(\tau) = \Phi_0 + \varepsilon\Phi_1 + \varepsilon^2\Phi_2 + \dots\,, \qquad q = 1 + q_2\varepsilon^2 + \dots$$

angesetzt, und in λ und q sind die Anteile nullter und erster Ordnung sogleich fortgelassen worden, da sie ersichtlich fortfallen werden. Aus (16) ergeben sich so zur Bestimmung von η über λ und Φ wieder rekursibel lösbare Gleichungen:

$$\Phi_0'' + \Phi_0 = 0\,, \qquad \Phi_1'' + \Phi_1 = -\beta_1 \cos\tau\Phi_0 - \sum_{j=2}^{N} \beta_j \cos(\alpha_j\tau - \psi_j)\,\Phi_0\,, \tag{16,1), (16,2}$$

$$\begin{aligned} \Phi_2'' + \Phi_2 = {} & -2q_2\Phi_0'' - 2D\Phi_0' - 2i\lambda_2\Phi_0' - \\ & - \beta_1 \cos\tau\Phi_1 - \sum_{j=2}^{N} \beta_j \cos(\alpha_j\tau - \psi_j)\,\Phi_1 - \tfrac{3}{2}\gamma A_0^2\Phi_0 - \tfrac{3}{2}\gamma A_0^2 \cos(2\tau - 2\vartheta_0)\,\Phi_0\,. \end{aligned} \tag{16,3}$$

Von

$$\Phi_0 = a_0 \cos(\tau - b_0)$$

als Lösung von (16,0) mit den Integrationskonstanten a_0 und b_0 ausgehend, erhält man über

$$\begin{aligned} \Phi_1 = {} & -\frac{\beta_1 a_0}{2} \cos b_0 + \frac{\beta_1 a_0}{6} \cos(2\tau - b_0) - \sum_{j=2}^{N} \frac{\beta_j a_0}{2} \frac{\cos(\alpha_j\tau - \tau - \psi_j + b_0)}{-(\alpha_j - 1)^2 + 1} - \\ & - \sum_{j=2}^{N} \frac{\beta_j a_0}{2} \frac{\cos(\alpha_j\tau + \tau - \psi_j - b_0)}{-(\alpha_j + 1)^2 + 1} \end{aligned}$$

aus (16,1) nun in (16,2)

$$\Phi_2'' + \Phi_2 = R_1 \cos(\tau - b_0) + R_2 \sin(\tau - b_0) + \ldots$$

erstmals Terme, die durch $R_1 = R_2 = 0$ beseitigt werden müssen, um auch Φ_2 in der gewünschten Weise fast-periodisch berechnen zu können. Dazu erscheint eine große Fülle weiterer Terme, die aber in (16,2) nicht zu Säkulargliedern Veranlassung geben und daher nicht niedergeschrieben werden sollen, da die Rechnung hier mit der ersten Näherung λ_2 abgebrochen werden kann. Aus $R_1 = 0$ folgt

$$2i\lambda_2 = -2D + \frac{\beta_1^2}{4} \sin 2b_0 - \frac{3}{4}\gamma A_0^2 \sin(2b_0 - 2\vartheta_0), \tag{17,1}$$

und $R_2 = 0$ ergibt:

$$\frac{\beta_1^2}{4} \cos 2b_0 - \frac{3}{4}\gamma A_0^2 \cos(2b_0 - 2\vartheta_0) = -2q_0 + \frac{3}{2}\gamma A_0^2 - \frac{\beta_1^2}{6} - \sum_{j=2}^{N} \frac{\beta_j^2}{2(4-\alpha_j^2)}. \tag{17,2}$$

In (17,1) und (17,2) ist mit $a_0 \neq 0$ die triviale Lösung ausgeschlossen worden, und mit A_0 und ϑ_0 gehen Amplitude und Phase der gegebenen, auf Stabilität zu prüfenden Lösung (12,0) ein. — Um zunächst (12,0) für $\vartheta_0 = 0$ zu testen, erhält man für diesen Fall aus (17,1) und (17,2) durch Elimination der Integrationskonstanten b_0 den Anteil λ_2 des charakteristischen Exponenten zu

$$2i\lambda_2 = -2D \pm \sqrt{\left(\frac{\beta_1^2}{4} - \frac{3}{4}\gamma A_0^2\right)^2 - \left(-2q_2 + \frac{3}{2}\gamma A_0^2 - \frac{\beta_1^2}{6} - \sum_{j=2}^{N}\frac{\beta_j^2}{2(4-\alpha_j)^2}\right)^2}, \tag{18}$$

der in der hier durchgeführten Näherung über Stabilität entscheidet. Für den hinsichtlich der Stabilität ungünstigsten Fall ohne Dämpfung ($D = 0$) ergeben sich für verschwindende Werte der Wurzel die beiden Grenzkurven

$$q = 1 + \left(-\frac{5}{24}\beta_1^2 - \sum_{j=2}^{N}\frac{\beta_j^2}{4(4-\alpha_j^2)} + \frac{9}{8}\gamma A_0^2\right)\varepsilon^2 + \ldots, \tag{19}$$

$$q = 1 + \left(\frac{1}{24}\beta_1^2 - \sum_{j=2}^{N}\frac{\beta_j^2}{4(4-\alpha_j^2)} + \frac{3}{8}\gamma A_0^2\right)\varepsilon^2 + \ldots, \tag{20}$$

die einen Instabilitätsbereich einschließen. Offenbar liegt die zu $D = 0$, $\vartheta_0 = 0$ von (12,0) gehörende Frequenzkurve (15) mit

$$q = 1 + \left(-\frac{5}{24}\beta_1^2 - \sum_{j=2}^{N}\frac{\beta_j^2}{4(4-\alpha_j^2)} + \frac{3}{8}\gamma A_0^2\right)\varepsilon^2 + \ldots \tag{21}$$

ganz außerhalb der durch (19) und (20) gegebenen Grenzen, und die zu (21) gehörende Schwingung (12,0) $y_0 = A_0 \times \cos\tau$ ist daher in der hier durchgeführten Näherung stabil. Eine entsprechende Rechnung für $\vartheta_0 = \pi/2$ in (12,0) ergibt den durch

$$q = 1 + \left(-\frac{5}{24}\beta_1^2 - \sum_{j=2}^{N}\frac{\beta_j^2}{4(4-\alpha_j^2)} + \frac{3}{8}\gamma A_0^2\right)\varepsilon^2 + \ldots \tag{22}$$

$$q = 1 + \left(\frac{1}{24}\beta_1^2 - \sum_{j=2}^{N}\frac{\beta_j^2}{4(4-\alpha_j^2)} + \frac{9}{8}\gamma A_0^2\right)\varepsilon^2 + \ldots \tag{23}$$

gegebenen Instabilitätsbereich. Die zu $\vartheta_0 = \pi/2$ gehörende Lösung ist nach (12,0) nun durch $y_0 = A_0 \sin\tau$ gegeben, hat für $D = 0$ die Frequenzgleichung

$$q = 1 + \left(\frac{1}{24}\beta_1^2 - \sum_{j=2}^{N}\frac{\beta_j^2}{4(4-\alpha_j^2)} + \frac{3}{8}\gamma A_0^2\right)\varepsilon^2 + \ldots \tag{24}$$

und liegt ganz innerhalb der Grenzen (22), (23), so daß die zu prüfende Lösung als instabil keine physikalische Realität hat.

Doppelte Eigenfrequenz in der Nähe der Summe zweier Parameterfrequenzen

Interessanterweise gibt es noch weitere Instabilitätsbereiche der Ruhelage, die allerdings auch nur eine geringe Frequenzbreite proprotional zu ε^2 haben. So findet man in der Nähe von $2\omega_e = \omega_1 + \omega_i = \omega(1 + \alpha_i)$, $i = 2, 3, \ldots, N$ solche Bereiche. Sie dürfen nicht mit den Instabilitäten verwechselt werden, die man von Schwingungssystemen mit mehreren Freiheitsgraden und mit harmonischer Parametererregung her kennt. Bei ihnen gibt es nur die eine Parameterfrequenz ω, und diese bildet mit zwei Systemeigenfrequenzen (zum Beispiel ω_{e1} und ω_{ei}) einen Instabilitätsbereich bei $2\omega = \omega_{e1} + \omega_{e2}$ [6]. Deren Frequenzbreite ist allerdings von der Größenordnung ε, und diese andere Art von Instabilitätsbereichen bei harmonischer Parametererregung ist somit technisch wichtiger. Bei der hier betrachteten fast-periodischen Parametererregung eines Einzelschwingers sind aber wieder nur Effekte der geringeren Größenordnung ε^2 zu erwarten, so daß auch wieder die Größenordnungsvoraussetzungen der Gleichungen (11,0) bis (11,2) zugrundegelegt werden. Mit hier

$$q = \frac{\omega(\varepsilon = 0)}{\omega_e} = \frac{2}{1 + \alpha_i} \tag{25}$$

nach (2) und (4) wird unverändert vorausgesetzt, daß alle Parameterfrequenzen einander proportional sind. In nullter Näherung erhält man mit den Integrationskonstanten A_0 und ϑ_0 mit nun q_0 nach (25) aus (11,0)

$$y_0 = A_0 \cos\left(\frac{1}{2}\tau + \frac{\alpha_i}{2}\tau - \vartheta_0\right), \tag{26}$$

wenn man wieder von der Ruhelage $A_0 = 0$ absieht. Aus (11,1) folgt weiter

$$y_1 = -\frac{\beta_1 A_0}{2}\frac{\cos\left(\frac{1}{2}\tau - \frac{\alpha_i}{2}\tau + \vartheta_0\right)}{-q_0^2\left(\frac{1}{2}-\frac{\alpha_i}{2}\right)^2+1} - \frac{\beta_1 A_0}{2}\frac{\cos\left(\frac{3}{2}\tau + \frac{\alpha_i}{2}\tau - \vartheta_0\right)}{-q_0^2\left(\frac{3}{2}+\frac{\alpha_i}{2}\right)^2+1} -$$

$$-\sum_{l=2}^{N}\frac{\beta_l A_0}{2}\frac{\cos\left(\alpha_l\tau - \frac{1}{2}\tau - \frac{\alpha_i}{2}\tau - \psi_l + \vartheta_0\right)}{-q_0^2\left(\alpha_l - \frac{1}{2}-\frac{\alpha_i}{2}\right)^2+1} - \sum_{l=2}^{N}\frac{\beta_l A_0}{2}\frac{\cos\left(\alpha_l\tau + \frac{1}{2}\tau + \frac{\alpha_i}{2}\tau - \psi_l - \vartheta_0\right)}{-q_0^2\left(\alpha_l + \frac{1}{2}+\frac{\alpha_i}{2}\right)^2+1} \tag{27}$$

und damit entstehen in (11,2) nach elementarer, jedoch längerer Rechnung wieder Terme, die zu Säkulargliedern Veranlassung geben, und viele weitere Terme, die jedoch nicht niedergeschrieben werden sollen:

$$q_0^2 y_2'' + y_2 = R_1 \cos\left(\frac{1}{2}\tau + \frac{\alpha_i}{2}\tau - \vartheta_0\right) + R_2 \sin\left(\frac{1}{2}\tau + \frac{\alpha_i}{2}\tau - \vartheta_0\right) + \dots .$$

Für q_0 nach (25) müssen also R_2 und R_1 verschwinden und diese Koeffizienten liefern so die Gleichungen:

$$2D - \frac{\beta_1\beta_i(1+\alpha_i)^2}{8\alpha_i}\sin(2\vartheta_0 - \psi_i) = 0 \tag{28}$$

und

$$q_2 = -\frac{\beta_1^2(1+\alpha_i)}{8\alpha_i(2+\alpha_i)} - \sum_{j=2}^{N}\frac{\beta_j^2(1+\alpha_j)}{8[(1+\alpha_i)^2-\alpha_j^2]} - \frac{\beta_1\beta_i(1+\alpha_i)}{8\alpha_i}\cos(2\vartheta_0 - \psi_i) + \frac{3}{4}\frac{\gamma A_0^2}{(1+\alpha_i)}. \tag{29}$$

Hieraus läßt sich wieder leicht die Phase $(2\vartheta_0 - \psi_i)$ eliminieren, so daß man einfacher

$$q = \frac{2}{1+\alpha_i} + \left(-\frac{\beta_1^2(1+\alpha_i)}{8\alpha_i(2+\alpha_i)} - \sum_{j=2}^{N}\frac{\beta_j^2(1+\alpha_i)}{8[(1+\alpha_i)^2-\alpha_j^2]} \mp\right.$$

$$\left.\mp\sqrt{\frac{\beta_1\beta_i(1+\alpha_i)^2}{64\alpha_i^2} - \frac{4D^2}{(1+\alpha_i)^2}} + \frac{3}{4}\frac{\gamma A_0^2}{1+\alpha_i}\right)\varepsilon^2 + \dots \tag{30}$$

erhält. Für den Grenzfall verschwindender Dämpfung ist $D = 0$, und (28) verlangt für die Integrationskonstante ϑ_0 abhängig von der gegebenen Phase ψ_i entweder

$$\vartheta_0 = \frac{\psi_i}{2} \quad \text{mit} \quad y_0 = A_0 \cos\left(\frac{1}{2}\tau + \frac{\alpha_i}{2}\tau - \frac{\psi_i}{2}\right) \tag{31}$$

oder anders

$$\vartheta_0 = \frac{\psi_i}{2} \pm \frac{\pi}{2} \quad \text{mit} \quad y_0 = \pm A_0 \sin\left(\frac{1}{2}\tau + \frac{\alpha_i}{2}\tau - \frac{\psi_i}{2}\right). \tag{32}$$

Die zugehörige Frequenz-Amplituden-Gleichung ist von der Form

$$q = \frac{2}{1+\alpha_i} + \left(-\frac{\beta_1^2(1+\alpha_i)}{8\alpha_i(2+\alpha_i)} - \sum_{j=2}^{N}\frac{\beta_j^2(1+\alpha_i)}{8[(1+\alpha_i)^2-\alpha_j^2]} \mp \frac{\beta_1\beta_i(1+\alpha_i)}{8\alpha_i} + \frac{3}{4}\frac{\gamma A_0^2}{1+\alpha_i}\right)\varepsilon^2 + \dots, \tag{33}$$

worin das obere Vorzeichen für (31), das untere für (32) gilt. Um zu erkennen, welche der beiden zu (33) gehörenden Schwingungen als stabil für die instabile Ruhelage eintritt, hat man die lineare Störungsgleichung (16) hier mit dem Ansatz (17) für $q_0 = 2/1 + \alpha_i$ zu lösen. Dies führt von $\Phi_0 = a_0 \cos\left(\frac{\tau}{q_0} - b_0\right)$ ausgehend in schon mehrfach beschriebener Weise auf

$$2iq_0\lambda_2 = -2D + \frac{\beta_1\beta_i(1+\alpha_i)^2}{8\alpha_i}\sin(2b_0 - \psi_i) - \frac{3}{4}\gamma A_0^2 \sin(2b_0 - 2\vartheta_0) \tag{34}$$

und

$$q = \frac{2}{1+\alpha_i} + \left(-\frac{\beta_1^2(1+\alpha_i)}{8\alpha_i(2+\alpha_i)} - \sum_{j=2}^{N}\frac{\beta_j^2(1+\alpha_i)}{8[(1+\alpha_i)^2-\alpha_j^2]} - \frac{\beta_1\beta_i(1+\alpha_i)}{8\alpha_i}\cos(2b_0 - \psi_i) + \right.$$

$$\left.+\frac{3}{2}\frac{\gamma A_0^2}{1+\alpha_i} + \frac{3}{4}\frac{\gamma A_0^2}{1+\alpha_i}\cos(2b_0 - 2\vartheta_0)\right)\varepsilon^2 + \dots . \tag{35}$$

Für verschwindende Dämpfung ($D = 0$) werde die Stabilitätsprüfung zunächst für (31) geleistet. Dafür ist also $\vartheta_0 = \psi_i/2$, und in (34) muß über die Integrationskonstante b_0 in der Störung so verfügt werden, daß $\sin(2b_0 - \psi_i) = 0$ ist, um die Grenzen des Instabilitätsbereiches zu finden. Für den Grenzfall $D = 0$ gibt es dann die beiden Möglichkeiten $\cos(2b_0 - \psi_i) = 1$ und $\cos(2b_0 - \psi_i) = -1$, und so entstehen durch das Doppelvorzeichen und die zugehörigen beiden verschiedenen Zahlenwerte bei der über γ eingehenden Nichtlinearität die Bereichsgrenzen

$$q = \frac{2}{1+\alpha_i} + \left(- \frac{\beta_1^2(1+\alpha_i)}{8\alpha_i(2+\alpha_i)} - \sum_{j=2}^{N} \frac{\beta_j^2(1+\alpha_i)}{8[(1+\alpha_i)^2 - \alpha_j^2]} \mp \frac{\beta_1\beta_i(1+\alpha_i)}{8\alpha_i} + \begin{Bmatrix} \frac{9}{4} \\ \frac{3}{4} \end{Bmatrix} \frac{\gamma A_0^2}{1+\alpha_i} \right) \varepsilon^2 + \dots \tag{36}$$

Ganz offenbar liegt aber die Frequenzkurve für den hier zunächst betrachteten Kosinus-Typ (31) mit den zugehörigen oberen Vorzeichen von (33) ganz außerhalb des durch (36) gegebenen Bereiches und ist damit instabil, denn hier wird ja stets eine überlineare Rückstellung betrachtet ($\gamma > 0$), und die Amplituden β_j der Parametererregung sind wegen des Ansatzes mit einer Phase ψ_j in (1) ihrer Natur nach stets positiv.

Für den Sinus-Typ (32) erhält man entsprechend (36) nun

$$q = \frac{2}{1+\alpha_i} + \left(- \frac{\beta_1^2(1+\alpha_i)}{8\alpha_i(2+\alpha_i)} - \sum_{j=2}^{N} \frac{\beta_j^2(1+\alpha_i)}{8[(1+\alpha_i)^2 - \alpha_j^2]} \mp \frac{\beta_1\beta_i(1+\alpha_i)}{8\alpha_i} + \begin{Bmatrix} \frac{3}{4} \\ \frac{9}{4} \end{Bmatrix} \frac{\gamma A_0^2}{1+\alpha_i} \right) \varepsilon^2 + \dots,$$

so daß nun anders die Frequenz-Gleichung (33) mit dem zugehörigen unteren Vorzeichen ganz innerhalb dieses Bereiches liegt und damit instabil ist. Erwartungsgemäß findet man also auch hier nur wieder eine stabile Schwingung, die ersatzweise für die instabile Ruhelage eintritt.

Doppelte Eigenfrequenz in der Nähe der Differenz zweier Parameterfrequenzen

Auch in der Nähe von $2\omega_e = -\omega_1 + \omega_i$ findet man bekanntlich Instabilitäten der Ruhelage, so daß auch bei dieser Abstimmlage ersatzweise nicht-lineare Schwingungen eintreten müssen. In der Tat liefert eine Rechnung, die gedanklich und formal ganz der vorstehenden entspricht, für $q_0 = 2/-1 + \alpha_i$ wieder den von $y_0 = A_0 \times \cos\left(-\frac{1}{2}\tau + \frac{\alpha_i}{2}\tau - \vartheta_0\right)$ ausgehenden Kosinus-Typ als ersatzweise für die Ruhelage stabile, nicht-lineare parametererregte Schwingung. Auch für die Abstimmung $2\omega_e = \omega_1 - \omega_i$ mit $q_0 = 2/1 - \alpha_i$ erhält man gleiche Ergebnisse, denn dieser Frequenzansatz entspricht in (26) ja nur dem Übergang von der Phase ϑ_0 auf die Phase $-\vartheta_0$, und dies führt zu keinen mechanisch anderen Aussagen. Man kann also auch $2\omega_e = |-\omega_1 + \omega_i|$ schreiben.

4. Zusammenfassung

Auch bei fast-periodischer Parametererregung eines Schwingers kann seine Ruhelage bei gewissen Abstimmungen der Parametererregungsfrequenzen mit der Eigenfrequenz instabil werden, und es müssen dann ersatzweise nicht-lineare Schwingungen eintreten. Daher wurden zunächst nicht-lineare Lösungen mit Hilfe einer Störungsrechnung unter gewissen Kleinheitsvoraussetzungen näherungsweise berechnet, und in einer Stabilitätsbetrachtung ergab sich jeweils eine als stabil und damit als die gesuchte Schwingung. Bei fast-periodischer Parametererregung sind wie bei rein harmonischer Parametererregung vorzugsweise solche Schwingungen zu erwarten, wenn eine Parameterfrequenz ω_i mit der doppelten Eigenfrequenz ω_e zusammenfällt: $2\omega_e = \omega_i$, wie dies von der Mathieu-Gleichung her bekannt ist. Alle weiteren Instabilitätsbereiche haben mit ε^2 eine sehr viel geringere Frequenzbreite, und sie werden daher leichter von der Dämpfung ausgelöscht. Im Falle der Instabilität der Ruhelage muß ersatzweise aber auch bei diesen Abstimmungen eine nicht-lineare Schwingung eintreten. Man findet solche wieder in der von periodischer Parametererregung her bekannten Weise bei $\omega_e = \omega_i$, $(i = 1, \dots, N)$. Interessanterweise gibt es aber auch solche Schwingungen beim Zusammenfallen der doppelten Eigenfrequenz mit der Summe oder der Differenz zweier Parametererregungsfrequenzen:

$$2\omega_e = |\pm \omega_i + \omega_j|, \quad (i, j = 1, \dots, N, i \neq j).$$

Literatur

1 Schmidt, G., Parametererregte Schwingungen, VEB Deutscher Verlag der Wissenschaften, Berlin 1975.
2 Schweitzer, G., Zur Stabilität eines parametererregten Schwingers, Z. Angew. Math. und Mech. **46** (1966) S. T 134–135.
3 Wedig, W., Regions of instability for a linear system with random parametric excitation, Lecture Notes in Mathematics, **294** (1972), S. 160–172.
4 Weidenhammer, F., Instabilitäten eines gedämpften Schwingers mit fast-periodischer Parametererregung, Ing.-Arch. **49** (1980) S. 187–193. (Hierin wird der Kehrwert von q verwendet und außer einigen Druckfehlern ist leider auch ein Vorzeichenfehler bei $\overset{(2)}{q}$ stehen geblieben.)
5 Weidenhammer, F., Nichtlineare Biegeschwingungen des axial pulsierend belasteten Stabes, Ing.-Archiv **20** (1952), S. 315–330.
6 Mettler, E., Allgemeine Theorie der Stabilität erzwungener Schwingungen elastischer Körper, Ing.-Arch. **17** (1949) S. 418–449.
7 Jacob M. Abel, On an almost periodic Mathieu equation, Quart. Journ. Appl. Mathematics, **28** (1970), S. 205–217.

Eingereicht am 24. 3. 1981

Anschrift: Prof. Dr. Fritz Weidenhammer, Bismarckstr. 11, 7500 Karlsruhe-1, BRD

ZAMM 61, 639 – 650 (1981)

W. A. Bassali / S. A. Obaid

On the Torsion of Elastic Cylindrical Bars

Certain trigonometric formulae and combinatorial identities are used to derive exact and closed expressions for the stress functions, torsional rigidities and peripheral shearing stresses of certain isotropic cylinders under torsion. The cross sections are bounded by the closed curves $r = a \left|\sin \frac{\theta}{n}\right|^n$ $(-\pi < \theta \leqq \pi)$, *where n is a positive integer* $(n > 1)$. *Complex variable methods are applied to obtain the complex torsion functions, torsional rigidities and shearing stresses when the sections are bounded by the contours* $r = a \cos^n \frac{\theta}{n} \left(-\frac{n\pi}{2} < \theta \leqq \frac{n\pi}{2}\right)$, *where n is a positive number* $(0 < n \leqq 2)$. *Numerical results are presented in the form of tables and graphs illustrating the variation of torsional rigidities and the distribution of tangential shearing stresses along the edges of the cross sections. The cardioid and lemniscate of Bernoulli are included as special cases.*

Zur Herleitung von exakten und geschlossenen Formeln für die Spannungsfunktionen, Torsionssteifigkeiten und peripheren Schubspannungen von bestimmten isotropen durch Torsion beanspruchten Zylindern werden trigonometrische Formeln und Kombinationsidentitäten benutzt. Die Querschnitte werden durch die geschlossenen Kurven $r = a \left|\sin \frac{\theta}{n}\right|^n$ $(-\pi < \theta \leqq \pi)$ *begrenzt, wobei n eine positive ganze Zahl* $(n > 1)$ *ist. Wenn die Querschnitte durch Kurven* $r = a \cos^n \frac{\theta}{n} \left(-\frac{n\pi}{2} < \theta \leqq \frac{n\pi}{2}\right)$ *begrenzt werden (n ist eine positive Zahl* $0 < n \leqq 2$), *ist die Methode der komplexen Veränderlichen angewendet worden, um die komplexen Torsionsfunktionen, Torsionssteifigkeiten und Schubspannungen zu erhalten. Numerische Ergebnisse werden in Form von Tabellen und Graphiken dargestellt, um die Veränderlichkeit der Torsionssteifigkeiten und die Verteilung der tangentialen Schubspannungen entlang den Querschnittsrändern zu illustrieren. Die Bernoullische Kardioide und Lemniskate sind als Spezialfälle enthalten.*

Для вывода точных и замкнутых выражений функций напряжений, жёсткостей при кручении и периферических напряжений сдвига определённых изотропных цилиндров при кручении используются тригонометрические формулы и комбинаторные тождества. Поперечные сечения ограничены замкнутыми кривыми $r = a \left|\sin \frac{\theta}{n}\right|^n$ $(-\pi < \theta \leqq \pi)$, причём, n-целые положительные числа $(n > 1)$. Если поперечное сечение ограничено кривыми $r = a \cos^n \frac{\theta}{n} \left(-\frac{n\pi}{2} < \theta \leqq \frac{n\pi}{2}\right)$, где n-положительное число $(0 < n \leqq 2)$, для определения комплексных функций кручения, жёсткостей при кручении и напряжений сдвига использован метод комплексного переменного. Численные результаты представлены в форме таблиц и графиков, демонстрирующих изменение жёсткостей при кручении и распределение касательных напряжений сдвига вдоль границ поперечного сечения. Кардиоида и лемниската Бернулли представляют собой частный случай.

1. Introduction

The classical torsion problem has long been a favorite of the physicist and of the applied mathematician. Real and complex variable methods have been applied to the problem for various forms of boundary including polygons, angles, sectors, limacons, lemniscates and others; references are to be found in standard textbooks and comprehensive reviews mentioned in previous papers by one of the authors [2], [3], [4] and in a recent paper by the authors [5]. Sections bounded by the systems of closed curves

$$r^m = a^m \cos n\theta\,, \qquad r^m = a^m \cos n\theta/(1 + \lambda \cos 2\theta)\,, \qquad m + n = 2\,, \qquad 0 \leqq n < 2\,, \qquad |\lambda| < 1\,, \quad (1)$$

have been treated in [5]. This paper deals with new solutions of the torsion problem corresponding to two families of closed sections bounded by curvilinear edges whose polar equations are

$$r = a \left|\sin \frac{\theta}{n}\right|^n, \qquad a > 0\,, \qquad (-\pi < \theta \leqq \pi)\,, \qquad n = 2, 3, 4, \ldots \quad (2a)$$

$$r = a \cos^n \frac{\theta}{n}, \qquad a > 0\,, \qquad \left(-\frac{n\pi}{2} < \theta \leqq \frac{n\pi}{2}\right), \qquad 0 < n \leqq 2\,. \quad (2b)$$

Denoting these boundaries by Γ_n and C_n, respectively, we see that C_1 is a circle, Γ_2 and C_2 are cardioids while $C_{1/2}$ is the lemniscate of Bernoulli $r^2 = a^2 \cos 2\theta$. Figures are sketched showing the shapes of the sections corresponding to various values of the parameter n. Trigonometric formulae and combinatorial identities are used to deal with (2a), while conformal mapping is applied to investigate the torsion problem of sections bounded by (2b). Expressions for the torsional rigidity D_n corresponding to $n = 2, 3, 4$ of the sections bounded by Γ_n are computed while a general formula for D_n in the case of sections bounded by C_n is explicitly given in terms of gamma functions. Standard formulae are applied to study the variation of the tangential shearing stresses on the boundaries Γ_n and C_n and the distribution is illustrated graphically for certain values of n.

2. Mathematical formulation of the torsion problem

We consider the classical Saint-Vénant torsion of a homogeneous cylindrical bar of isotropic material and uniform cross section R bounded by a simple closed curve C in the z-plane where $z = x + iy = r\,e^{i\theta}$. The z-plane is chosen to coincide with the plane $Z = 0$ perpendicular to the generators of the cylinder. Twisting of the bar by two equal and opposite torques M about its axis applied at its two bases leads to the displacements (u, v, w) and the non-vanishing stresses $\widehat{xZ}$, $\widehat{yZ}$ which are given [7, p. 596] by

$$u = -\tau y Z\,, \qquad v = \tau x Z\,, \qquad w = \tau\varphi(x, y) = \tfrac{1}{2}\tau\,[\Omega(z) + \bar{\Omega}(\bar{z})]\,, \tag{3}$$

$$\widehat{xZ} - i\widehat{yZ} = \mu\tau[\Omega'(z) - i\bar{z}]\,, \tag{4}$$

where τ is the constant twist per unit length, μ is the modulus of rigidity of the material and

$$\Omega(z) = \varphi(r, \theta) + i\psi(r, \theta) \tag{5}$$

is the complex torsion function which satisfies the boundary condition

$$\psi(r, \theta) = \tfrac{1}{2}r^2 \quad \text{or} \quad \Omega(z) - \bar{\Omega}(\bar{z}) = iz\bar{z} \quad \text{along } C\,. \tag{6}$$

The stress function $\Psi(r, \theta)$ is defined by

$$\Psi(r, \theta) = \psi(r, \theta) - \tfrac{1}{2}r^2\,. \tag{7}$$

It satisfies the conditions of finiteness and continuity in R,

$$\nabla^2\Psi(r, \theta) = -2 \quad \text{in } R\,, \qquad \Psi(r, \theta) = 0 \quad \text{on } C\,. \tag{8}$$

The torsional rigidity D is the ratio of the torque M to the twist and is furnished by

$$D = M/\tau = 2\mu \iint_R \Psi(r, \theta)\, r\, dr\, d\theta\,, \tag{9}$$

where the integration is carried over the cross section.

The radial and transverse components of stress are given by

$$\widehat{rZ} = \frac{\mu\tau}{r}\frac{\partial\Psi}{\partial\theta}\,, \qquad \widehat{\theta Z} = -\mu\tau\frac{\partial\Psi}{\partial r}\,. \tag{10}$$

We assume that the conformal transformation

$$z = z(\zeta) \qquad (z = x + iy = r\,e^{i\theta}, \zeta = \xi + i\eta = \varrho\,e^{i\psi}) \tag{11}$$

maps the region R on the unit disc $|\zeta| \leqq 1$. Curvilinear coordinates (ϱ, ψ) are thus introduced into the z-plane. The complex torsion function can be expressed in terms of ζ and we have

$$\Omega(z) = \omega(\zeta)\,, \tag{12}$$

where $\omega(\zeta)$ is regular inside the unit circle γ. It is known [7, p. 600] that the torsional rigidity is given by the formula

$$D = \mu(I + J)\,, \tag{13}$$

where I is the polar moment of inertia for the cross section and

$$J = \tfrac{1}{2}\int_\gamma z(\sigma)\,\bar{z}(\sigma^{-1})\,\omega'(\sigma)\,d\sigma \qquad (\sigma = e^{i\psi})\,. \tag{14}$$

The shearing stresses $\widehat{\varrho Z}$ and $\widehat{\psi Z}$ at any point of R are determined by the convenient formula [7, p. 602]

$$\widehat{\varrho Z} - i\widehat{\psi Z} = \frac{\mu\tau\zeta}{\varrho|z'(\zeta)|}[\omega'(\zeta) - iz'(\zeta)\,\bar{z}(\bar{\zeta})]\,. \tag{15}$$

Since $(\widehat{\varrho Z})_{\varrho=1} = 0$ it follows that the peripheral shear stress at any point of C is

$$S = (\widehat{\psi Z})_{\varrho=1} = \frac{\mu\tau\sigma}{|z'(\sigma)|}[z'(\sigma)\,\bar{z}(\sigma^{-1}) + i\omega'(\sigma)]\,, \tag{16}$$

where $\sigma = e^{i\psi}$ is the corresponding point on γ.

3. Solution of the torsion problem corresponding to the first family of closed sections

In this article we consider sections bounded by the contours Γ_n with polar equation

$$r = a\left|\sin\frac{\theta}{n}\right|^n \qquad (-\pi < \theta \leqq \pi)\,, \qquad a > 0\,, \tag{17}$$

where n is a positive integer $(n > 1)$. The contour Γ_n is symmetric with respect to the initial line and has a cusp at the pole O which corresponds to $\theta = 0$. The curves Γ_n corresponding to some values of n are plotted in Fig. 1. Assuming that P is any point on Γ_n, it is easily shown that the angle between the radius OP and the tangent to the curve at P equals $\frac{\theta}{n}$. The two tangents to Γ_n at the point $Q\left(\pi, a\sin^n\frac{\pi}{n}\right)$ include an angle $\frac{2\pi}{n}\,(n > 1)$.

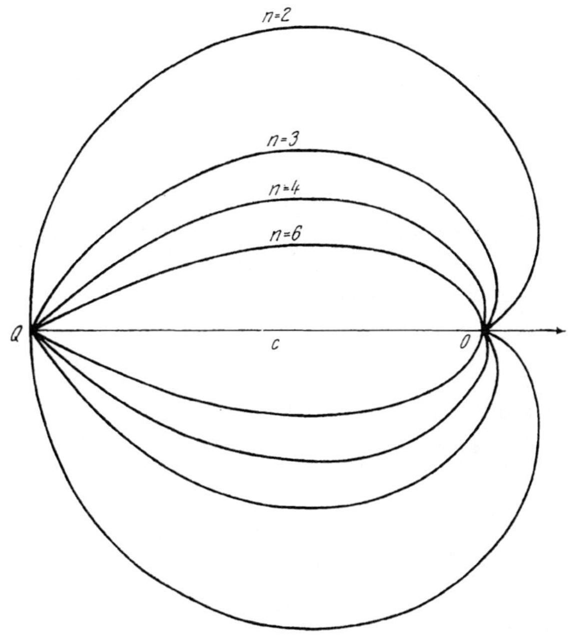

Fig. 1. Cross sections bounded by the system of curves

$r = c\left|\sin\frac{\theta}{n}\Big/\sin\frac{\pi}{n}\right|^n \; (-\pi < \theta \leqq \pi)\,, \qquad n = 2, 3, 4, \ldots$

We now proceed to derive an exact expression in finite terms involving fractional powers of r for the stress function $\Psi(r, \theta)$ at any point $P(r, \theta)$ inside or on $\Gamma_n(0 \leqq \theta \leqq \pi)$. Our starting point is to establish the following trigonometric formulae for even powers of $\cos x$ and $\sin x$: Assuming that

$$c_\nu = 2\cos \nu x \qquad (\nu = 1, 2, 3, \ldots)\,, \tag{18}$$

we are going to prove that

$$c_1^{2n} = \sum_{\nu=1}^{n} p_\nu^n c_1^\nu c_\nu\,, \tag{19}$$

where

$$p_\nu^n = \binom{2n-\nu-1}{n-1} \qquad (\nu \leqq n) \tag{20}$$

are the binomial coefficients. Writing $t = e^{ix}$ we see that the right side of (19) equals

$$\sum_{\nu=1}^{n} p_\nu^n (t + t^{-1})^\nu (t^\nu + t^{-\nu}) = \sum_{\nu=1}^{n} p_\nu^n \{(t^2+1)^\nu + (t^{-2}+1)^\nu\}\,. \tag{21}$$

The constant term in the binomial expansion of (21) in powers of t is $2\sum_{\nu=1}^{n} p_\nu^n$ and the coefficient of t^{2n-2m} or t^{2m-2n} $(0 \leqq m < n)$ is $\sum_{\nu=1}^{n} p_\nu^n \binom{\nu}{n-m}$. In Appendix I we prove the two combinatorial identities

$$2\sum_{\nu=1}^{n} p_\nu^n = \binom{2n}{n}, \qquad \sum_{\nu=1}^{n} p_\nu^n \binom{\nu}{n-m} = \binom{2n}{m} \qquad (0 \leqq m < n)\,. \tag{22}$$

Thus the right side of (19) equals

$$\binom{2n}{n} + \sum_{m=0}^{n-1} \binom{2n}{m} (t^{2n-2m} + t^{2m-2n}) = (t + t^{-1})^{2n} = c_1^{2n}\,.$$

Another proof of (19) is to apply the following formulae [6, p. 471] for c_ν in terms of powers of c_1:

$$c_{2n+1} = (2n+1)(-1)^n \sum_{\nu=0}^{n} \frac{(-1)^\nu}{n+\nu+1} \binom{n+\nu+1}{2\nu+1} c_1^{2\nu+1}\,, \tag{23a}$$

$$c_{2n} = 2n(-1)^n \sum_{\nu=0}^{n} \frac{(-1)^\nu}{n+\nu} \binom{n+\nu}{2\nu} c_1^{2\nu}\,. \tag{23b}$$

Introducing these values in the right side of (19) and transforming the finite double sum to a single finite sum in powers of c_1 we find that the coefficient of c_1^{2n} is unity while the coefficient of any lower power vanishes upon using the combinatorial identity

$$\sum_{\nu=0}^{r} (-1)^\nu (n-\nu) \binom{n+\nu-1}{n-1} \binom{n-r}{r-\nu} = 0 \qquad (n > r)\,, \tag{24}$$

which is proved in Appendix I.

Setting $\pi/2 - x$ instead of x in (19) and adopting the notation

$$s_\nu = 2\sin \nu x \qquad (\nu = 1, 2, 3, \ldots)\,, \tag{25}$$

we arrive at the two formulae

$$s_1^{2n} = (-1)^{(n-1)/2}\left[\sum_{\nu=0}^{(n-1)/2}(-1)^\nu\binom{n+2\nu-1}{n-1}s_1^{n-2\nu}s_{n-2\nu} + \sum_{\nu=1}^{(n-1)/2}(-1)^{\nu-1}\binom{n+2\nu-2}{n-1}s_1^{n-2\nu+1}c_{n-2\nu+1}\right] (n \text{ odd}), \tag{26a}$$

$$s_1^{2n} = (-1)^{n/2}\left[\sum_{\nu=0}^{(n/2)-1}(-1)^\nu\binom{n+2\nu-1}{n-1}s_1^{n-2\nu}c_{n-2\nu} + \sum_{\nu=1}^{n/2}(-1)^\nu\binom{n+2\nu-2}{n-1}s_1^{n-2\nu+1}s_{n-2\nu+1}\right] (n \text{ even}). \tag{26b}$$

Assuming that $P(r, \theta)$ is any point on Γ_n $(0 \leqq \theta \leqq \pi)$ then equation (2a) gives

$$\sin\frac{\theta}{n} = \left(\frac{r}{a}\right)^{1/n} \qquad (0 \leqq \theta \leqq \pi). \tag{27}$$

Putting θ/n instead of x in (26) and using (27) we deduce that on the upper half of Γ_n

$$\frac{r^2}{2a^2} = \frac{(-1)^{(n-1)/2}}{2^n}\left[\sum_{\nu=0}^{(n-1)/2}\frac{(-1)^\nu}{2^{2\nu}}\binom{n+2\nu-1}{n-1}\left(\frac{r}{a}\right)^{1-(2\nu/n)}\sin\left(1-\frac{2\nu}{n}\right)\theta + \sum_{\nu=1}^{(n-1)/2}\frac{(-1)^{\nu-1}}{2^{2\nu-1}}\binom{n+2\nu-2}{n-1}\left(\frac{r}{a}\right)^{1-(2\nu-1)/n}\cos\left(1-\frac{2\nu-1}{n}\right)\theta\right] (n \text{ odd}), \tag{28a}$$

$$\frac{r^2}{2a^2} = \frac{(-1)^{n/2}}{2^n}\left[\sum_{\nu=0}^{(n/2)-1}\frac{(-1)^\nu}{2^{2\nu}}\binom{n+2\nu-1}{n-1}\left(\frac{r}{a}\right)^{1-(2\nu/n)}\cos\left(1-\frac{2\nu}{n}\right)\theta + \sum_{\nu=1}^{n/2}\frac{(-1)^\nu}{2^{2\nu-1}}\binom{n+2\nu-2}{n-1}\left(\frac{r}{a}\right)^{1-(2\nu-1)/n}\sin\left(1-\frac{2\nu-1}{n}\right)\theta\right] (n \text{ even}). \tag{28b}$$

For any point $P(z = r\,e^{i\theta})$ inside or on Γ_n $(0 \leqq \theta \leqq \pi)$ we now define the stress function as

$$\Psi_n(r,\theta) = \frac{a^2}{2^n}\left[-2^{n-1}\frac{r^2}{a^2} + (-1)^{(n-1)/2}\left\{\sum_{\nu=0}^{(n-1)/2}\frac{(-1)^\nu}{2^{2\nu}}\binom{n+2\nu-1}{n-1}\left(\frac{r}{a}\right)^{1-(2\nu/n)}\sin\left(1-\frac{2\nu}{n}\right)\theta + \sum_{\nu=1}^{(n-1)/2}\frac{(-1)^{\nu-1}}{2^{2\nu-1}}\binom{n+2\nu-2}{n-1}\left(\frac{r}{a}\right)^{1-(2\nu-1)/n}\cos\left(1-\frac{2\nu-1}{n}\right)\theta\right\}\right], \tag{29a}$$

$$\Psi_n(r,\theta) = \frac{a^2}{2^n}\left[-2^{n-1}\frac{r^2}{a^2} + (-1)^{n/2}\left\{\sum_{\nu=0}^{(n/2)-1}\frac{(-1)^\nu}{2^{2\nu}}\binom{n+2\nu-1}{n-1}\left(\frac{r}{a}\right)^{1-(2\nu/n)}\cos\left(1-\frac{2\nu}{n}\right)\theta + \sum_{\nu=1}^{n/2}\frac{(-1)^\nu}{2^{2\nu-1}}\binom{n+2\nu-2}{n-1}\left(\frac{r}{a}\right)^{1-(2\nu-1)/n}\sin\left(1-\frac{2\nu-1}{n}\right)\theta\right\}\right]. \tag{29b}$$

according as n is odd or even, respectively. The value of the stress function Ψ_n at the conjugate point $P'(\bar{z} = r\,e^{-i\theta})$ is defined as equal to its value at $P(z = r\,e^{i\theta})$. This is in agreement with the reflection principle [1, p. 170]. It is clear that the expression between the curly brackets in (29a) or (29b) is the sum of a finite number of terms of the form $r^k \frac{\cos}{\sin} k\theta$ which are known to be harmonic functions of (r, θ). Thus the expressions (29 a, b) for $\Psi_n(r, \theta)$ are finite, continuous and satisfy Poisson's equation $\nabla^2\Psi = -2$ at any point inside Γ_n. Furthermore, equations (28 a, b) show that these expressions vanish on Γ_n. The two conditions (8) are therefore satisfied and the formulae (29 a, b) yield the complete solution of the torsion problem for cross sections bounded by Γ_n.

We now give explicit expressions for $\Psi_2(r, \theta)$, $\Psi_3(r, \theta)$, $\Psi_4(r, \theta)$ and insert them in (9) to evaluate D_2, D_3 and D_4. Equations (29 a, b) lead to

$$\Psi_2(r,\theta) = \frac{1}{4}a\left(-\frac{2r^2}{a} - r\cos\theta + a^{1/2}r^{1/2}\sin\frac{1}{2}\theta\right), \tag{30a}$$

$$\Psi_3(r,\theta) = \frac{1}{8}a\left(-\frac{4r^2}{a} - r\sin\theta - \frac{3}{2}a^{1/3}r^{2/3}\cos\frac{2}{3}\theta + \frac{3}{2}a^{2/3}r^{1/3}\sin\frac{\theta}{3}\right), \tag{30b}$$

$$\Psi_4(r,\theta) = \frac{1}{16}a\left(-\frac{8r^2}{a} + r\cos\theta - 2a^{1/4}r^{3/4}\sin\frac{3}{4}\theta - \frac{5}{2}a^{1/2}r^{1/2}\cos\frac{\theta}{2} + \frac{5}{2}a^{3/4}r^{1/4}\sin\frac{\theta}{4}\right). \tag{30c}$$

After trigonometric manipulation and extensive computations we obtain

$$D_2 = 2\mu a^4\int_0^{\pi/2}\left(\frac{1}{15}\sin^6\varphi + \frac{1}{6}\sin^8\varphi\right)d\varphi = \frac{17}{256}\pi\mu a^4, \tag{31a}$$

$$D_3 = \frac{3}{2}\mu a^4\int_0^{\pi/3}\left(\frac{9}{112}\sin^8\varphi + \frac{1}{8}\sin^{10}\varphi + \frac{1}{3}\sin^{12}\varphi\right)d\varphi = \frac{243}{2048}\left(\frac{131}{243}\pi - \frac{487}{560}\sqrt{3}\right)\mu a^4, \tag{31b}$$

$$D_4 = \mu a^4 \int_0^{\pi/4} \left(\frac{1}{9}\sin^{10}\varphi + \frac{5}{33}\sin^{12}\varphi + \frac{8}{33}\sin^{14}\varphi + \frac{2}{3}\sin^{16}\varphi\right) d\varphi = \frac{5}{8}\left(\frac{797}{8192}\pi - \frac{211}{693}\right)\mu a^4 . \tag{31c}$$

Writing the equation of Γ_n in the form

$$r = c\left|\sin\frac{\theta}{n}\Big/\sin\frac{\pi}{n}\right|^n, \qquad a = c\Big/\sin^n\frac{\pi}{n}, \tag{32}$$

we easily find that

$$D_2 = \frac{17}{256}\pi\mu c^4 = 0.2086\mu c^4 , \tag{33a}$$

$$D_3 = \frac{2}{3}\mu c^4\left(\frac{131}{243}\pi - \frac{487}{560}\sqrt{3}\right) = 0.1249\,\mu c^4 , \qquad D_4 = 10\,\mu c^4\left(\frac{797}{512}\pi - \frac{3376}{693}\right) = 0.1876\,\mu c^4 . \tag{33b), (33c}$$

Since the resultant shearing stress at any point of R attains its maximum value on the boundary C [7, p. 597] we now study the distribution of the tangential shearing stress S_n on Γ_n. If $\varphi = \theta/n$ is the angle between the radius OP at any point P on Γ_n and the tangent to Γ_n at P then

$$S_n = (\widehat{rZ})_{\Gamma_n}/\cos\varphi = \mu\tau\left(\frac{1}{r}\frac{\partial\Psi_n}{\partial\theta}\right)_{\Gamma_n}\Big/\cos\varphi . \tag{34}$$

Substitution for Ψ_n from (29 a, b) and for r from (2a) gives

$$S_n = \frac{\mu\tau a}{2^n\cos\varphi}(-1)^{(n-1)/2}\left[\sum_{\nu=0}^{(n-1)/2}(-1)^\nu\left(1-\frac{2\nu}{n}\right)\binom{n+2\nu-1}{n-1}\frac{\cos(n-2\nu)\varphi}{(2\sin\varphi)^{2\nu}} + \right.$$
$$\left. + \sum_{\nu=1}^{(n-1)/2}(-1)^\nu\left(1-\frac{2\nu-1}{n}\right)\binom{n+2\nu-2}{n-1}\frac{\sin(n-2\nu+1)\varphi}{(2\sin\varphi)^{2\nu-1}}\right] \quad (n \text{ odd}), \tag{35a}$$

$$S_n = \frac{\mu\tau a}{2^n\cos\varphi}(-1)^{n/2}\left[\sum_{\nu=0}^{(n/2)-1}(-1)^{\nu+1}\left(1-\frac{2\nu}{n}\right)\binom{n+2\nu-1}{n-1}\frac{\sin(n-2\nu)\varphi}{(2\sin\varphi)^{2\nu}} + \right.$$
$$\left. + \sum_{\nu=1}^{n/2}(-1)^\nu\left(1-\frac{2\nu-1}{n}\right)\binom{n+2\nu-2}{n-1}\frac{\cos(n-2\nu+1)\varphi}{(2\sin\varphi)^{2\nu-1}}\right] \quad (n \text{ even}). \tag{35b}$$

With the notations (18) and (25), it is known [6, p. 471] that

$$s_{2n} = c_1\sum_{\nu=0}^{n-1}(-1)^\nu\binom{n+\nu}{2\nu+1}s_1^{2\nu+1}, \qquad c_{2n+1} = c_1\sum_{\nu=0}^{n}(-1)^\nu\binom{n+\nu}{2\nu}s_1^{2\nu}. \tag{36a), (36b}$$

Introducing these expressions in (35 a, b) and using the following combinatorial identity

$$\sum_{\nu=0}^{n-r}(-1)^\nu(2n-2r+1-\nu)\binom{n-r}{\nu}\binom{2r+\nu-2}{n-1} = (-1)^{n-r}\frac{n}{r}\binom{2r-2}{r-1} \quad (n \geqq r), \tag{37}$$

which is proved in Appendix I we find, after considerable algebraic manipulation to transform the double sum into a single sum, that whether n is odd or even the tangential shearing stress S_n at any point (r, θ) of Γ_n is furnished by the simple formula

$$S_n(\theta) = \mu\tau a\left(\sin\frac{\theta}{n}\right)^n\sum_{\nu=0}^{n-1}\frac{\binom{2\nu}{\nu}}{(\nu+1)\left(2\sin\frac{\theta}{n}\right)^{2\nu+1}} \qquad (0 \leqq \theta \leqq \pi). \tag{38}$$

Hence we have

$$S_2(\theta)/\mu\tau a = \frac{1}{2}\sin\frac{\theta}{2} + \frac{1}{8}\operatorname{cosec}\frac{\theta}{2}, \qquad S_3(\theta)/\mu\tau a = \frac{1}{2}\sin^2\frac{\theta}{3} + \frac{1}{8} + \frac{1}{16}\operatorname{cosec}^2\frac{\theta}{3}, \tag{39a), (39b}$$

$$S_4(\theta)/\mu\tau a = \frac{1}{2}\sin^3\frac{\theta}{4} + \frac{1}{8}\sin\frac{\theta}{4} + \frac{1}{16}\operatorname{cosec}\frac{\theta}{4} + \frac{5}{128}\operatorname{cosec}^3\frac{\theta}{4}. \tag{39c}$$

Table 1

θ	$S_2/\mu\tau a$	$S_3/\mu\tau a$	$S_4/\mu\tau a$	θ	$S_2/\mu\tau a$	$S_3/\mu\tau a$	$S_4/\mu\tau a$
10°	1.4778	18.6134	472.1155	100°	0.5462	0.4830	0.7560
20°	0.8067	4.7691	59.7313	110°	0.5622	0.4786	0.6391
30°	0.6124	2.2128	18.0621	120°	0.5774	0.4829	0.5625
40°	0.5365	1.3268	7.8444	130°	0.5911	0.4932	0.5129
50°	0.5071	0.9259	4.1735	140°	0.6029	0.5077	0.4820
60°	0.5000	0.7178	2.5356	150°	0.6124	0.5249	0.4647
70°	0.5047	0.6018	1.6956	160°	0.6193	0.5438	0.4575
80°	0.5159	0.5360	1.2218	170°	0.6236	0.5636	0.4578
90°	0.5303	0.5000	0.9362	180°	0.6250	0.5833	0.4640

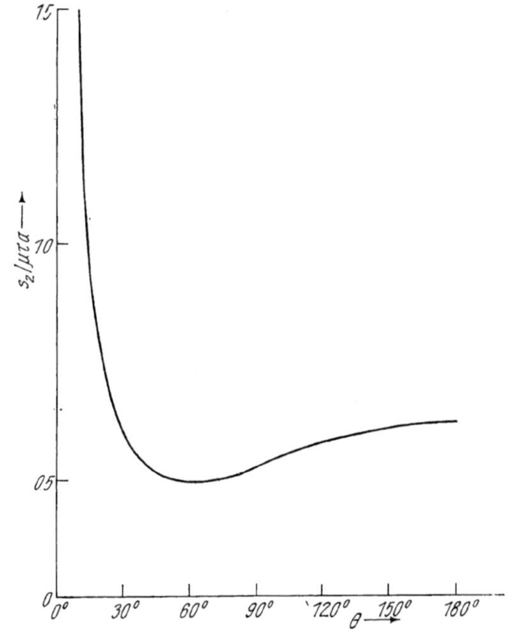

Fig. 2. Variation of peripheral shearing stress along the cardioid $r = a \sin^2 \frac{\theta}{2} \quad (-\pi < \theta \leq \pi)$

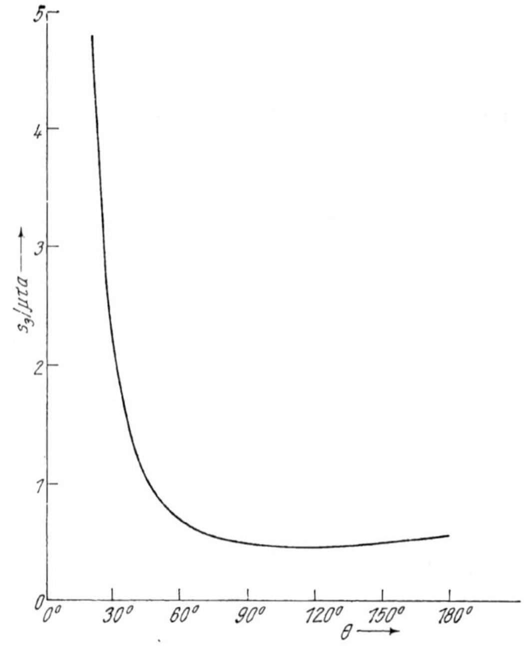

Fig. 3. Variation of peripheral shearing stress along the curve $r = a \left|\sin^3 \frac{\theta}{3}\right| \quad (-\pi < \theta \leq \pi)$

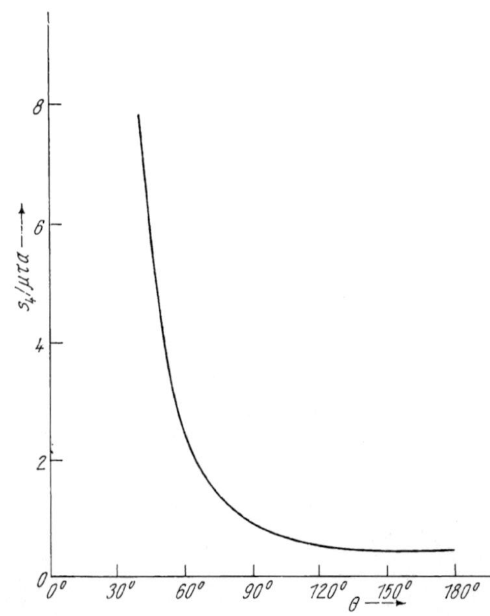

Fig. 4. Variation of peripheral shearing stress along the curve $r = a \sin^4 \frac{\theta}{4} \; (-\pi < \theta \leq \pi)$

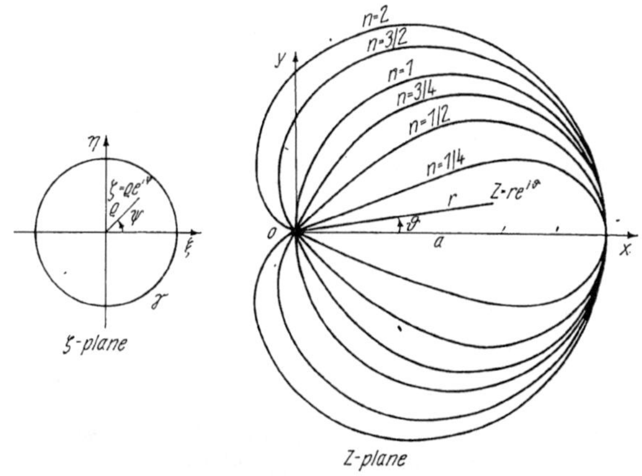

Fig. 5. Mapping of cross sections bounded by the system of curves $r = a \cos^n \frac{\theta}{n} \left(-\frac{n\pi}{2} < \theta \leq \frac{n\pi}{2}\right)$ on the unit disc

The numerical values of these stresses are exhibited in Table 1 and graphically plotted in Figs. 2, 3 and 4, respectively.

4. Solution of the torsion problem corresponding to the second family of closed sections

In this article we apply Fourier series and complex variable methods to derive the complete solution of the torsion problem connected with the closed sections bounded by the contours C_n with polar equation

$$r = a \cos^n \frac{\theta}{n} \left(-\frac{n\pi}{2} < \theta \leq \frac{n\pi}{2}\right), \qquad a > 0, \tag{40}$$

where n is a real positive parameter subject to the restriction

$$0 < n \leq 2. \tag{41}$$

The shapes of the curves C_n corresponding to some values of n are sketched in Fig. 5. It is easily shown that the conformal transformation

$$z = c(1 + \zeta)^n, \qquad c = a/2^n \tag{42}$$

maps the region R inside C_n on $|\zeta| < 1$. If $z = r\,e^{i\theta}$ is the point on C_n corresponding to the point $\sigma = e^{i\psi}$ on the unit circle γ then $r\,e^{i\theta} = c(1 + \cos\psi + i\sin\psi)^n = c(2\cos\psi/2)^n\,e^{in\psi/2}$. Hence $r = c(2\cos\psi/2)^n$, $\theta = n\psi/2$ and elimination of ψ leads to (40).

It is worthy of mention that by rotating the initial line through an angle $\frac{1}{2}n\pi$ (putting $\theta = \Theta - \frac{1}{2}n\pi$) the curve (40) takes the form

$$r = a\sin^n\frac{\Theta}{n}, \qquad 0 \leqq \Theta \leqq n\pi, \qquad 0 < n \leqq 2.$$

In the special case $n = 1/m$ where m is a positive integer equation (40) becomes

$$r^m = a^m\cos m\theta = 2c^m\cos m\theta \qquad (-\pi/2m < \theta \leqq \pi/2m),$$

which is the polar equation of the generalised lemniscate. For $(-\pi < \theta \leqq \pi)$ we have m loops with m foci at the points

$$w_s = c\,e^{2s\pi i/m} \qquad (s = 0, 1, 2, \ldots, m-1).$$

If $z = r\,e^{i\theta}$ is any point on the loop $(-\pi/2m < \theta \leqq \pi/2m)$ and $z_s = z - w_s = r_s\,e^{is\theta_s}$ then it is easily shown that

$$r_1 r_2 \ldots r_m = c^m = a^m/2, \qquad \theta_1 + \theta_2 + \ldots + \theta_m = \psi = 2m\theta.$$

Our approach of deriving the complex torsion function corresponding to (40) is to find first the Fourier expansion of the function

$$f(\theta) = \cos^{2n}\frac{\theta}{n}\,(-L \leqq \theta \leqq L), \qquad L = \frac{1}{2}n\pi. \tag{43}$$

We have $f(\theta) = \frac{1}{2}A_0 + \sum_{k=1}^{\infty} A_k\cos\frac{k\pi\theta}{L}$, where

$$A_k = \frac{2}{L}\int_0^L f(\theta)\cos\frac{k\pi\theta}{L}\,d\theta = \frac{4}{n\pi}\int_0^{n\pi/2}\cos^{2n}\frac{\theta}{n}\cos\frac{2k\theta}{n}\,d\theta = \frac{4}{\pi}\int_0^{\pi/2}\cos^{2n}\varphi\cos 2k\varphi\,d\varphi \qquad (k = 0, 1, 2, \ldots). \tag{44}$$

The last integral can be evaluated by applying the formula [6, p. 9]

$$\int_0^{\pi/2}(\cos t)^\alpha\cos(\beta t)\,dt = \frac{\pi\Gamma(1+\alpha)\,2^{-\alpha-1}}{\Gamma(1+\frac{1}{2}\alpha+\frac{1}{2}\beta)\,\Gamma(1+\frac{1}{2}\alpha-\frac{1}{2}\beta)}, \qquad \operatorname{Re}\alpha > -1. \tag{45}$$

Hence

$$A_k = \frac{n}{2^{2n-2}}\cdot\frac{\Gamma(2n)}{\Gamma(1+n+k)\,\Gamma(1+n-k)} \qquad (k = 0, 1, 2, \ldots), \tag{46}$$

$$\cos^{2n}\frac{\theta}{n} = \frac{n\Gamma(2n)}{2^{2n-1}}\left[\frac{1}{n^2\Gamma^2(n)} + 2\sum_{k=1}^{\infty}\frac{\cos(2k\theta/n)}{\Gamma(1+n+k)\,\Gamma(1+n-k)}\right]. \tag{47}$$

Putting ψ instead of $2\theta/n$ it is easily seen that on C_n we have

$$\frac{1}{2}r^2 = nc^2\,\Gamma(2n)\operatorname{Re}\left[\frac{1}{n^2\Gamma^2(n)} + 2\sum_{k=1}^{\infty}\frac{\sigma^k}{\Gamma(1+n+k)\,\Gamma(1+n-k)}\right]. \tag{48}$$

The complex torsion function $\omega(\zeta)$ at any point ζ inside or on γ may now be taken as

$$\omega(\zeta) = nic^2\,\Gamma(2n)\left[\frac{1}{n^2\Gamma^2(n)} + 2\sum_{k=1}^{\infty}\frac{\zeta^k}{\Gamma(1+n+k)\,\Gamma(1+n-k)}\right]. \tag{49}$$

Setting $\zeta = (z/c)^{1/n} - 1$ we obtain

$$\Omega(z) = nic^2\Gamma(2n)\left[\frac{1}{n^2\Gamma^2(n)} + 2\sum_{k=1}^{\infty}\frac{\{(z/c)^{1/n} - 1\}^k}{\Gamma(1+n+k)\,\Gamma(1+n-k)}\right]. \tag{50}$$

For $0 < n \leqq 1$ the point $z = 0$ lies on C_n, $\Omega'(0) = 0$ and $\Omega(z)$ is analytic at all points inside or on C_n. For $1 < n \leqq 2$ the point $z = 0$ lies on C_n and $\Omega'(0) = \infty$. For values of $n > 2$ the point $z = 0$ lies inside C_n, $\Omega'(0) = \infty$ and $\Omega(z)$ becomes not analytic inside C_n. Hence we consider only the range $0 < n \leqq 2$.

We now apply (13) to evaluate the torsional rigidity D_n. The polar moment of inertia I is given by

$$I = 2\int_0^{n\pi/2}\int_0^{a\cos^n\theta/n} r^3\,dr\,d\theta = \frac{1}{2}na^4\int_0^{\pi/2}\cos^{4n}\varphi\,d\varphi = \frac{1}{4}\pi c^4\,\frac{\Gamma(4n)}{\Gamma^2(2n)}. \tag{51}$$

This value can also be obtained by applying the formula [7, p. 601]

$$I = -\frac{1}{4}i\int_{C_n}\bar{z}^2 z\,dz. \tag{52}$$

Substituting from (42) and (49) in (14) we get

$$J = nc^4 \Gamma(2n) \sum_{k=1}^{\infty} \frac{k L_k}{\Gamma(1+n+k)\,\Gamma(1+n-k)}, \tag{53}$$

where

$$L_k = i \int_\gamma (1+\sigma)^{2n} \sigma^{k-n-1}\, d\sigma . \tag{54}$$

Setting $\sigma = e^{i\psi}$, $d\sigma = i\, e^{i\psi}\, d\psi$ we obtain

$$L_k = -2^{2n+1} \int_0^{\pi} \left(\cos\frac{\psi}{2}\right)^{2n} \cos k\psi\, d\psi = -2^{2n+2} \int_0^{\pi/2} \cos^{2n}\varphi \cos 2k\varphi\, d\varphi = -\frac{4\pi n\, \Gamma(2n)}{\Gamma(1+n+k)\,\Gamma(1+n-k)} \tag{55}$$

upon applying (45). Introducing this value in (53) yields

$$J = -4\pi\, n^2 c^4 \Gamma^2(2n) \sum_{k=1}^{\infty} \frac{k}{\Gamma^2(1+n+k)\,\Gamma^2(1+n-k)}. \tag{56}$$

It is proved in Appendix II that the sum of the infinite series in (56) is $1/\{4n^3\Gamma^4(n)\}$. Hence we have

$$J = -\frac{\pi c^4}{n} \frac{\Gamma^2(2n)}{\Gamma^4(n)}. \tag{57}$$

Substitution from (51) and (57) in (13) leads to the simple formula

$$D_n = \pi\mu c^4 \left[\frac{\Gamma(4n)}{4\Gamma^2(2n)} - \frac{\Gamma^2(2n)}{n\Gamma^4(n)}\right] = \frac{\pi\mu a^4}{2^{4n}} \left[\frac{\Gamma(4n)}{4\Gamma^2(2n)} - \frac{\Gamma^2(2n)}{n\Gamma^4(n)}\right]. \tag{58}$$

Numerical values of $D_n/\mu a^4$ and $D_n/\mu c^4$ are listed in Table 2 and plotted in Figs. 6, 7.

Table 2

n	$D_n/\mu a^4$	$D_n/\mu c^4$	n	$D_n/\mu a^4$	$D_n/\mu c^4$
0.1	0.0014	0.0018	1.1	0.1101	2.3252
0.2	0.0068	0.0118	1.2	0.1219	3.3959
0.3	0.0153	0.0352	1.3	0.1335	4.9058
0.4	0.0257	0.0780	1.4	0.1448	7.0238
0.5	0.0372	0.1488	1.5	0.1559	9.9807
0.6	0.0492	0.2597	1.6	0.1669	14.0927
0.7	0.0615	0.4281	1.7	0.1776	19.7905
0.8	0.0738	0.6781	1.8	0.1881	27.6618
0.9	0.0860	1.0433	1.9	0.1985	38.5054
1	0.0982	1.5708	2	0.2086	53.4070

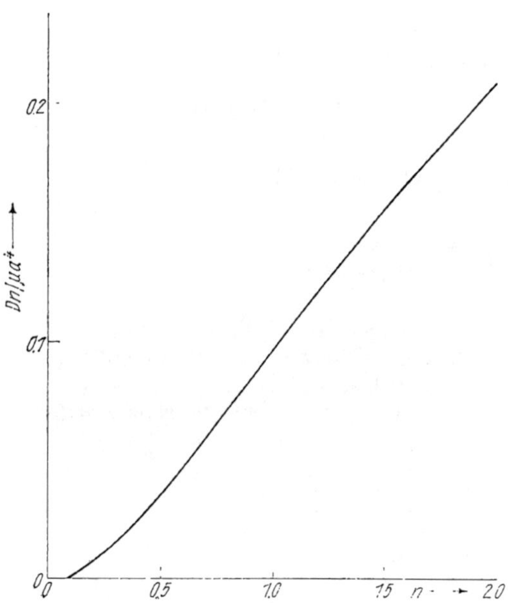

Fig. 6. Variation of $D_n/\mu a^4$ with n for cross sections bounded by the contours $r = a \cos^n \frac{\theta}{n}$ $(0 < n \leq 2)$

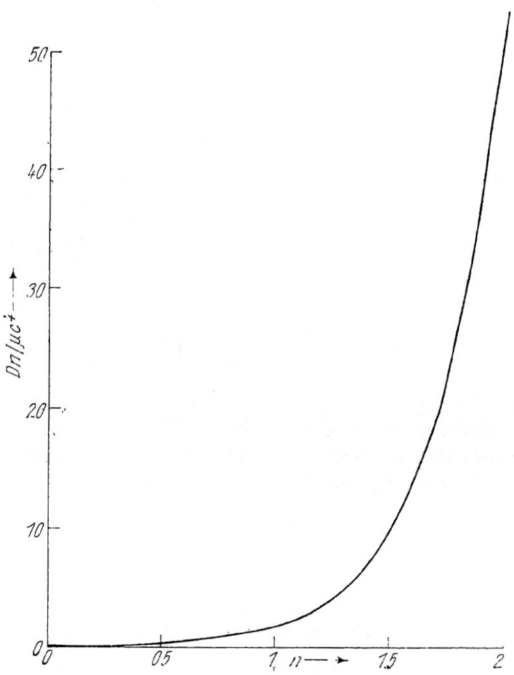

Fig. 7. Variation of $D_n/\mu c^4$ with n for cross sections bounded by the contours $r = c\left(2\cos\frac{\theta}{n}\right)^n$ $(0 < n \leq 2)$

The components of shearing stress at any point of the cross section can now be computed by applying (15). Direct substitution from (42) and (49) gives

$$\widehat{\varrho Z} = \mu\tau c(1+\varrho^2+2\varrho\cos\psi)^{(1-n)/2}\left[(1+\varrho^2+2\varrho\cos\psi)^{n-1}\sin\psi - 2\Gamma(2n)\sum_{k=1}^{\infty}\frac{k\varrho^{k-1}\sin k\psi}{\Gamma(1+n+k)\Gamma(1+n-k)}\right], \tag{59a}$$

$$\widehat{\psi Z} = \mu\tau c(1+\varrho^2+2\varrho\cos\psi)^{(1-n)/2}\left[(1+\varrho^2+2\varrho\cos\psi)^{n-1}(\varrho+\cos\psi) - 2\Gamma(2n)\sum_{k=1}^{\infty}\frac{k\varrho^{k-1}\cos k\psi}{\Gamma(1+n+k)\,\Gamma(1+n-k)}\right]. \tag{59b}$$

Setting $\varrho = 1$ we find that the expression between the square brackets in (59a) becomes

$$\sin\psi\left(4\cos^2\frac{\psi}{2}\right)^{n-1} - 2\Gamma(2n)\sum_{k=1}^{\infty}\frac{k\sin k\psi}{\Gamma(1+n+k)\,\Gamma(1+n-k)}. \tag{60}$$

Putting $\theta/n = \psi/2$ in (47) and then differentiating both sides with respect to ψ we see that the expression (60) vanishes. This proves that $(\widehat{\varrho Z})_{\varrho=1} = 0$ as expected.

The tangential shearing stress at any point P of C_n is

$$S_n(\psi) = (\widehat{\psi Z})_{\varrho=1} = \frac{\mu\tau a}{2^{1+n}}\left(2\cos\frac{\psi}{2}\right)^{1-n}\left[\left(2\cos\frac{\psi}{2}\right)^{2n} - 4\Gamma(2n)\sum_{k=1}^{\infty}\frac{k\cos k\psi}{\Gamma(1+n+k)\,\Gamma(1+n-k)}\right]. \tag{61}$$

This result can be directly achieved by using (16). Applying the formula

$$\Gamma(k-n)\,\Gamma(1+n-k) = \frac{(-1)^{k+1}\pi}{\sin n\pi} \qquad (n \neq 1, 2), \tag{62}$$

we can write (61) in the form

$$S_n(\psi) = \frac{\mu\tau a}{2^{1+n}}\left(2\cos\frac{\psi}{2}\right)^{1-n}\left[\left(2\cos\frac{\psi}{2}\right)^{2n} + \frac{4}{\pi}\Gamma(2n)\sin n\pi\sum_{k=1}^{\infty}\frac{(-1)^k k\Gamma(k-n)\cos k\psi}{\Gamma(1+n+k)}\right] \qquad (n \neq 1, 2). \tag{63}$$

For $n = \frac{1}{2}, \frac{3}{2}$ the infinite series appearing in (63) is summed in Appendix III and we arrive at the closed forms

$$S_{1/2}(\theta) = \mu\tau a\sqrt{\cos 2\theta}\left[\cos 2\theta + \frac{1}{\pi}\left\{\sin 2\theta\ln\frac{1+\sin 2\theta}{1-\sin 2\theta} - 2\right\}\right], \tag{64a}$$

$$S_{3/2}(\theta) = \mu\tau a\sqrt{\sec\frac{2\theta}{3}}\left[\cos^3\frac{2\theta}{3} - \frac{1}{\pi}\left\{\frac{1}{3} + \cos\frac{4\theta}{3} - \sin\frac{2\theta}{3}\cos^2\frac{2\theta}{3}\ln\frac{1+\sin(2\theta/3)}{1-\sin(2\theta/3)}\right\}\right]. \tag{64b}$$

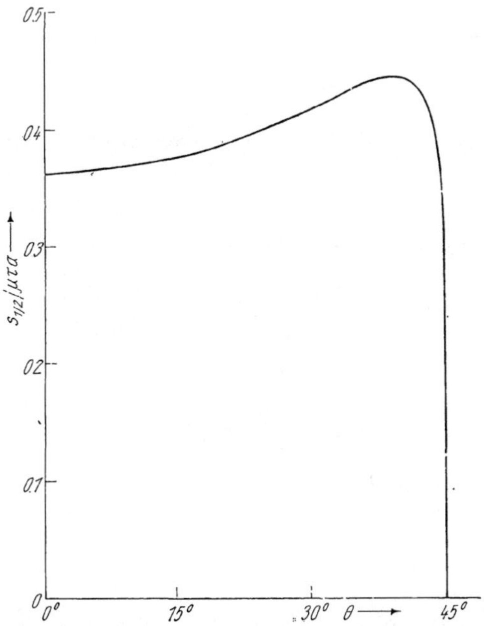

Fig. 8. Distribution of shearing stress along the lemniscate of Bernoulli $r^2 = a^2\cos 2\theta$

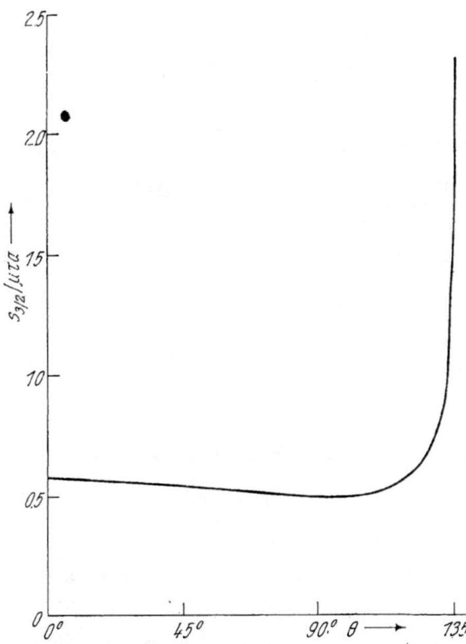

Fig. 9. Distribution of shearing stress along the contour $r^2 = a^2\cos^3\frac{2\theta}{3}$

For $n = 2$ the expression (61) reduces to

$$S_2(\theta) = \mu\tau a\left(\frac{1}{2}\cos\frac{\theta}{2} + \frac{1}{8}\sec\frac{\theta}{2}\right); \tag{64c}$$

which is in agreement with (39a). Table 3 and Figs. 8, 9 and 2 show the distribution of the peripheral shear in these three cases.

Table 3

ψ	$S_{1/2}/\mu\tau a$	$S_{3/2}/\mu\tau a$	$S_2/\mu\tau a$	ψ	$S_{1/2}/\mu\tau a$	$S_{3/2}/\mu\tau a$	$S_2/\mu\tau a$
0°	0.3634	0.5756	0.6250	140°	0.4349	0.5116	0.5365
10°	0.3638	0.5750	0.6236	150°	0.4421	0.5315	0.6124
20°	0.3648	0.5731	0.6193	160°	0.4436	0.5863	0.8067
30°	0.3665	0.5700	0.6124	170°	0.4241	0.7558	1.4778
40°	0.3690	0.5657	0.6029	171°	0.4191	0.7905	1.6324
50°	0.3722	0.5603	0.5911	172°	0.4130	0.8323	1.8268
60°	0.3762	0.5538	0.5774	173°	0.4055	0.8836	2.0781
70°	0.3810	0.5465	0.5622	174°	0.3961	0.9483	2.4146
80°	0.3865	0.5385	0.5462	175°	0.3842	1.0328	2.8875
90°	0.3929	0.5302	0.5303	176°	0.3687	1.1488	3.5992
100°	0.4001	0.5219	0.5159	177°	0.3477	1.3207	4.7883
110°	0.4081	0.5143	0.5047	178°	0.3169	1.6118	7.1711
120°	0.4168	0.5085	0.5000	179°	0.2645	2.2739	14.3285
130°	0.4259	0.5064	0.5071	180°	0	∞	∞

Appendix I

If m, n are positive integers, $n > m$ then equations (1) and (2), p. 7 of Riordan [9] state that

$$\sum_{k=0}^{n-m}\binom{n-1-k}{m-1} = \binom{n}{m}, \qquad \sum_{k=0}^{n-m}\binom{n-j-k}{m-k}\binom{j+k-1}{j-1} = \binom{n}{m}. \tag{1), (2}$$

Replacing n by $2n$ and putting $m = n$ in (1) we get

$$\sum_{k=0}^{n}\binom{2n-1-k}{n-1} = \binom{2n}{n}. \qquad \text{Hence} \qquad \binom{2n-1}{n-1} + \sum_{k=1}^{n}\binom{2n-1-k}{n-1} = \binom{2n}{n}.$$

Noting that $\binom{2n-1}{n-1} = \frac{1}{2}\binom{2n}{n}$ we deduce that

$$2\sum_{k=1}^{n}\binom{2n-k-1}{n-1} = \binom{2n}{n}. \tag{3}$$

The identity (2) may be rewritten in the form

$$\sum_{k=0}^{m}\binom{n-j-k}{n-j-m}\binom{j+k-1}{j-1} = \binom{n}{m}. \tag{4}$$

Putting $2n$ instead of n in (4) and setting $j + k = \nu + 1$, $j + m = n + 1$ we find

$$\sum_{\nu=n-m}^{n}\binom{2n-\nu-1}{n-1}\binom{\nu}{n-m} = \binom{2n}{m}. \tag{5}$$

With the usual convention $\binom{\nu}{s} = 0$ $(s > \nu)$ we see that the lower end of the range in (5) may be taken as $\nu = 1$ instead of $\nu = n - m$.

To prove the two identities (24) and (37) in the paper we consider the function

$$F(x, y) = (1 + x)^\alpha\,[y^m - (1 + x)^p]^\beta\, y^\gamma, \tag{6}$$

where α, β, γ are positive integers, m is any real number and $p = 1$ or -1. It is easily seen that

$$F(x, y) = \sum_{j=0}^{\beta}(-1)^j\binom{\beta}{j}(1 + x)^{\alpha+pj}\, y^{\gamma+m\beta-mj}, \qquad \left(\frac{\partial F}{\partial y}\right)_{y=1} = \sum_{j=0}^{\beta}(-1)^j\binom{\beta}{j}(\gamma + m\beta - mj)\,(1 + x)^{\alpha+pj}. \tag{7}$$

The coefficient of x^k in the binomial expansion of (7) is

$$\sum_{j=0}^{\beta}(-1)^j(\gamma + m\beta - mj)\binom{\beta}{j}\binom{\alpha + pj}{k}. \tag{8}$$

Differentiating (6) and then putting $y = 1$ we get

$$\left(\frac{\partial F}{\partial y}\right)_{y=1} = (1 + x)^\alpha\,[\gamma\{1 - (1 + x)^p\}^\beta + m\beta\,\{1 - (1 + x)^p\}^{\beta-1}]. \tag{9}$$

We now consider the two cases

(i): if $p = 1$ then the expression (9) reduces to

$$(-1)^{\beta-1}\,(1 + x)^\alpha\,(m\beta x^{\beta-1} - \gamma x^\beta) \tag{10}$$

and equating (8) to the coefficient of x^k in the binomial expansion of (10) we obtain the combinatorial identity

$$\sum_{j=0}^{\beta}(-1)^j(\gamma + m\beta - mj)\binom{\beta}{j}\binom{\alpha + j}{k} = (-1)^{\beta-1}\left\{m\beta\binom{\alpha}{k - \beta + 1} - \gamma\binom{\alpha}{k - \beta}\right\}. \tag{11}$$

Taking $m = 1, \alpha = 2r - 2, \beta = n - r, \gamma = n - r + 1, k = n - 1$ and simplifying the right side we obtain equation (37) in the paper.

(ii): if $p = -1$ then (9) simplifies to

$$\gamma x^\beta (1 + x)^{\alpha - \beta} + m\beta x^{\beta - 1}(1 + x)^{\alpha - \beta + 1} . \tag{12}$$

Equating (8) to the coefficient of x^k in the binomial expansion of (12) we find

$$\sum_{j=0}^{\beta} (-1)^j (\gamma + m\beta - mj) \binom{\beta}{j} \binom{\alpha - j}{k} = \gamma \binom{\alpha - \beta}{k - \beta} + m\beta \binom{\alpha - \beta + 1}{\alpha - k} . \tag{13}$$

Taking $m = -1, \alpha = n + r - 1, \beta = n - r, \gamma - \beta = n - r, k = n - 1$ we get

$$\sum_{j=0}^{n-r} (-1)^j (n - r + j) \binom{n + r - j - 1}{r - j} \binom{n - r}{j} = 0 \quad (n > r) . \tag{14}$$

It is easily shown that the upper end of the range in (14) may be replaced by r instead of $n - r$. Setting $r - j = \nu$ we obtain equation (24) in the paper.

Appendix II

It is required to prove that

$$H = \sum_{k=1}^{\infty} \frac{k}{\Gamma^2(1 + n + k)\, \Gamma^2(1 + n - k)} = \frac{1}{4n^3 \Gamma^4(n)} .$$

Using Pochhammer's symbol [6, p. 3]

$$(\alpha)_m = \alpha(\alpha + 1) \ldots (\alpha + m - 1) = \Gamma(\alpha + m)/\Gamma(\alpha) ,$$

we easily see that $k = k!/(k - 1)! = (2)_{k-1}/(k - 1)!$,

$$\Gamma(1 + n + k) = (1 + n)\,(2 + n)_{k-1}\, \Gamma(1 + n) ,$$

$$\Gamma(1 + n - k) = (-1)^k\, \Gamma(1 + n)/(-n)(-n + 1)_{k-1} .$$

Hence $H = \lambda \sum_{k=1}^{\infty} \left[\frac{(1 - n)_{k-1}}{(2 + n)_{k-1}}\right]^2 \frac{(2)_{k-1}}{(k - 1)!}$ where $\lambda = \frac{1}{n^2 (1 + n)^2\, \Gamma^4(n)}$.

Adopting the usual notation [8, p. 73], we have

$$H = \lambda \sum_{k=0}^{\infty} \left[\frac{(1 - n)_k}{(2 + n)_k}\right]^2 \frac{(2)_k}{(k)!} = \lambda\, {}_3F_2 \left[\begin{matrix} 2, 1 - n, 1 - n; \\ 2 + n, 2 + n \end{matrix} 1\right] .$$

Taking $a = 2, b = c = 1 - n$ in Dixon's formula [8, p. 92] we find that

$$H = \lambda \frac{(1 + n)^2}{4n} = \frac{1}{4n^3 \Gamma^4(n)} .$$

Appendix III

In this appendix we obtain in finite terms the sums of the two infinite series

$$H_n = \sum_{k=1}^{\infty} \frac{(-1)^k\, k \Gamma(k - n) \cos k\psi}{\Gamma(1 + n + k)} \qquad \left(n = \frac{1}{2}, \frac{3}{2}\right) .$$

Introducing the notation $J_m = \sum_{\nu=1}^{\infty} \frac{(-1)^\nu \cos \nu\psi}{2\nu + m}$ we see that

$$H_{1/2} = J_1 + J_{-1}, \quad H_{3/2} = \frac{1}{2} (J_3 + J_{-3} - J_1 - J_{-1}) .$$

Setting $t = e^{i\psi/2}$ and noticing that

$$\sum_{k=0}^{\infty} (-1)^k \frac{t^{2k+1}}{2k + 1} = \tan^{-1} t \; (|t| \leqq 1, t \neq \pm i) ,$$

we easily find that

$$J_1 = \operatorname{Re} (t^{-1} \tan^{-1} t) - 1 , \qquad J_{-1} = -\operatorname{Re} (t \tan^{-1} t) ,$$

$$J_3 = \cos \psi - \tfrac{1}{3} - \operatorname{Re} (t^{-3} \tan^{-1} t) , \qquad J_{-3} = \cos \psi + \operatorname{Re} (t^3 \tan^{-1} t) .$$

Applying the formula

$$\tan^{-1} e^{i\theta} = \frac{\pi}{4} + \frac{1}{4}\, i \ln \frac{1 + \sin \theta}{1 - \sin \theta} \qquad \left(-\frac{\pi}{2} < \theta < \frac{\pi}{2}\right)$$

we deduce that

$$H_{1/2} = J_1 + J_{-1} = \frac{1}{2} \sin \frac{\psi}{2} \ln \frac{1 + \sin \frac{\psi}{2}}{1 - \sin \frac{\psi}{2}} - 1 ,$$

$$J_3 + J_{-3} = 2 \cos \psi - \frac{1}{3} - \frac{1}{2} \sin \frac{3\psi}{2} \ln \frac{1 + \sin \frac{\psi}{2}}{1 - \sin \frac{\psi}{2}} ,$$

$$H_{3/2} = \frac{1}{3} + \cos \psi - \sin \psi \cos^2 \frac{\psi}{2} \ln \frac{1 + \sin \frac{\psi}{2}}{1 - \sin \frac{\psi}{2}} .$$

References

1 AHLFORS, L. V., Complex Analysis, An Introduction to the Theory of Analytic Functions of One Variable, 2nd Ed., McGraw-Hill, New York (1966).
2 BASSALI, W. A., Torsion of Beams whose Sections are Bounded by Certain Quartic Curves, J. Mech. Phys. Solids 7, 272–281 (1959).
3 BASSALI, W. A., The Torsion of Elastic Cylinders with Regular Curvilinear Cross Sections, J. Math. Phys. 38, 232–245 (1960).
4 BASSALI, W. A., The Classical Torsion Problem for Sections with Curvilinear Boundaries, J. Mech. Phys. Solids 8, 87–98 (1960).
5 BASSALI, W. A., and OBAID, S. A., Solutions of Saint-Venant's Torsion Problem for Certain Simply Connected Cross Sections, J. Univ. Kuwait (Sci.) 8 (1981).
6 MAGNUS, W., OBERHETTINGER, F., and SONI, R. P., Formulas and Theorems for the Special Functions of Mathematical Physics, 3rd. Ed., Springer-Verlag, New York (1966).
7 MUSKHELISHVILI, N. I., Some Basic Problems of the Mathematical Theory of Elasticity, 4th Ed., Noordhoof International Publishing Company, Leyden, The Netherlands (1954).
8 RAINVILLE, E. D., Special Functions, Chelsea Publishing Company, New York (1960).
9 RIORDAN, J., Combinatorial Identities, John Wiley and Sons, New York (1968).

Eingereicht am 8. 10. 1980

Anschrift: Dr. W. A. BASSALI, Prof., Faculty of Science, Kuwait University, P.O. Box 5969, Kuwait

ZAMM 61, 651–656 (1981)

H. Stolarski

Objective Strain Acceleration Measures in Nonlinear Analysis of Structures

A method of derivation of the accelerations of generalized strains for various structures is presented. Contrary to the common approach to the problem, consisting in successive differentiation of the components of appropriate strain tensor, the method utilizes the general formulas, defining objective time derivatives of tensors. In the structural analysis, such a way of derivation of the higher rates of the strain tensor is less involved and permits the rates associated with various objective time derivatives to be systematically obtained.

Es wird eine Methode zur Herleitung der Beschleunigungen von verallgemeinerten Beanspruchungen für verschiedene Strukturen präsentiert. Im Gegensatz zur üblichen Behandlung des Problems, wo die Komponenten eines geeigneten Beanspruchungstensors aufeinanderfolgend differenziert werden, benutzt die Methode allgemeine Formeln, die objektive Zeitableitungen eines Tensors definieren. Solch eine Herleitung der höheren Raten des Beanspruchungstensors ist bei der Untersuchung der Kontinua weniger verwickelt und erlaubt, die Raten, die den objektiven Zeitableitungen entsprechen, in einer systematischen Weise zu erhalten.

Дан метод определения ускорений общих деформаций для различных структурных сред. Напротив обычного решения этой проблемы, где последовательно дифференцируется нужный тензор деформации, этот метод пользуется общими формулами, которые определяют объективные производные во времени тензоров. При исследовании структурных сред такой путь вывода высших скоростей тензора деформации прост и позволяет систематически определить эти скорости, которые соответствуют различным объективным производным во времени.

1. Introduction

Objective time derivative of tensors have been introduced in nonlinear continuum mechanics primarily to properly describe changes of the physical quantities related to those tensors. Each of them vanishes if the considered continuum undergoes rigid body motion only. This is the main property of the derivatives.

Some of the objective derivatives, like Zaremba-Jaumann [4, 14] and Truesdell [13] derivatives, have been developed to describe the stress rates (see also [8]); other, like Rivlin-Ericksen derivative [9] (see also [1]), to describe the strain rates. There are also derivatives introduced formally (without any reference to a physical quantity) so as to assure their independence of rigid body motion.

The kinematical approach to the analysis of structural problems requires the general expressions, defining objective time derivatives of the strain tensor to be transformed into relationships describing appropriate generalized quantities. Because of a frequent use of the velocity strain tensor the transformations for the derivatives of the first order are fairly known. However, until the works [10] and [11], there had been no indication of practical need for the second order objective rates of the strain tensor. Consequently, there is no paper presenting derivation of the generalized strain acceleration. In addition to that, considerations regarding validity of an extremum principle in nonlinear dynamics of rigid-plastic solids, [11, 12], indicate that in some mechanical problems the objective time derivatives, developed originally for the stress analysis, can be appropriate to describe higher rates of the strain tensor. This fact justifies the need for the paper, presenting a unified method of derivation of the generalized strain accelerations, based on the definitions of various objective time derivatives.

In this paper it is shown how the accelerations of the generalized strains, appropriate for various structures, can be systematically derived from the expressions, defining the main objective derivatives of tensors. To accomplish the task the general formulas are transformed first to the form containing the acceleration and the velocity fields. Then the strain acceleration measures for arbitrary shells are derived within the Kirchhoff hypothesis. These are followed by specification leading to the shells of revolution and bars.

Although the use of the resulting formulas has so far been demonstrated in dynamics of plastic structures only, [10, 12], the kinematical considerations presented herein are selfcontained and are believed to be useful while solving other mechanical problems, requiring higher rates of the strain tensor.

2. Basic definitions

Considera convected coordinate system X^k, $k = 1, 2, 3$ imbedded into a moving body. By $\boldsymbol{q}_k^t$ and $\boldsymbol{q}_k^\tau$ denote the vector of the natural bases associated with the system at time t and τ respectively. The bases can be transformed into each other by making use of the relative deformation gradient $\boldsymbol{F}_{(t)}(\tau)$, [6]

$$\boldsymbol{q}_k^\tau = \boldsymbol{F}_{(t)}(\tau)\, \boldsymbol{q}_k^t \,. \tag{1a}$$

This implies that

$$\boldsymbol{q}_k^t = \boldsymbol{F}_{(t)}^{-1}(\tau)\, \boldsymbol{q}_k^\tau \,, \qquad \boldsymbol{q}_t^k = \boldsymbol{F}_{(t)}^T(\tau)\, \boldsymbol{q}_\tau^k \,, \qquad \boldsymbol{q}_\tau^k = \left(\boldsymbol{F}_{(t)}^{-1}(\tau)\right)^T \boldsymbol{q}_t^k \,, \tag{1b}$$

where $\boldsymbol{q}_t^k$ and $\boldsymbol{q}_\tau^k$ are the bases dual to $\boldsymbol{q}_k^t$ and $\boldsymbol{q}_k^\tau$.

According to the polar decomposition theorem, [6],

$$\boldsymbol{F}_{(t)}(\tau) = \boldsymbol{R}_{(t)}(\tau)\, \boldsymbol{U}_{(t)}(\tau) \,, \tag{2}$$

where $U_{(t)}(\tau)$ is symmetric and positive definite, $R_{(t)}(\tau)$ is orthogonal and

$$F_{(t)}(t) = R_{(t)}(t) = U_{(t)}(t) = \mathbf{1}\,, \tag{3}$$

where $\mathbf{1}$ is the unit, second order tensor.

In the next part of the work the following derivatives of the above tensors will be used

$$L \equiv \frac{\partial}{\partial \tau}\left(F_{(t)}(\tau)\right)_{\tau=t}, \qquad D \equiv \frac{\partial}{\partial \tau}\left(U_{(t)}(\tau)\right)_{\tau=t}, \qquad W \equiv \frac{\partial}{\partial \tau}\left(R_{(t)}(\tau)\right)_{\tau=t}. \tag{4}$$

By virtue of Eqs. (2) and (3)

$$L = D + W\,, \qquad D = \operatorname{sym} L\,, \qquad W = \operatorname{ant} L\,, \tag{5a}$$

$$\frac{\partial}{\partial \tau}\left(F_{(t)}^{-1}(\tau)\right)_{\tau=t} = -L\,, \qquad \frac{\partial}{\partial \tau}\left(R_{(t)}^{-1}(\tau)\right) = -W = W^T\,. \tag{5b}$$

Further evaluation of the Eq. $(4)_1$ leads to the conclusion that, [3, 6]

$$L \equiv \nabla v \equiv v_{,k} \otimes q^k\,, \tag{6}$$

where ∇v is the spatial gradient of the current velocity field v at time t, $\otimes$ stands for the tensor product.

The tensor D describes the first rate of the strain tensor (i.e. the strain velocity) referred to the current configuration. The second rate (the strain accelerations) will be obtained by objective differentiation of the strain velocity tensor D. To indicate that the further evaluations apply to a class of the objective derivatives considered are Zaremba-Jaumann [4, 14], Rivlin-Ericksen [9], Oldroyd [7], and Truesdell [13] time derivatives. Their use will make it evident that other derivatives can be incorporated into the same formula.

All the derivatives mentioned above can be described by a single formula

$$\overset{A}{D} \equiv \frac{\partial}{\partial \tau}\left(A_{(t)}^T(\tau)\, D(\tau)\, A_{(t)}(\tau)\right)_{\tau=t} \tag{7}$$

in which $\overset{A}{D}$ stands for the resulting tensor, associated with both D and A. Specifing the tensor $A_{(t)}(\tau)$ and taking into account the Eqs. (3) and (4) one can obtain the following derivatives:

i) Zaremba-Jaumann derivative

$$A_{(t)}(\tau) \equiv R_{(t)}(\tau) \Rightarrow \overset{\circ}{D} = \dot{D} - WD + DW\,, \tag{8}$$

ii) Rivlin-Ericksen derivative

$$A_{(t)}(\tau) \equiv F_{(t)}(\tau) \Rightarrow \overset{\nabla}{D} = \dot{D} + L^T D + DL\,, \tag{9}$$

iii) Oldroyd derivative

$$A_{(t)}(\tau) \equiv \left(F_{(t)}^{-1}(\tau)\right)^T \Rightarrow \overset{\nabla_0}{D} = \dot{D} - LD - DL^T\,, \tag{10}$$

iv) Truesdell derivative

$$A_{(t)}(\tau) \equiv \sqrt{J}\left(F_{(t)}^{-1}(\tau)\right)^T \Rightarrow \overset{\nabla_T}{D} = \dot{D} - LD - DL^T + (\operatorname{tr} L)\, D\,. \tag{11}$$

In the last definition $J = \det F_{(t)}(\tau)$, whereas its derivative $\dot{J} = \operatorname{tr} L$, [6, 13]. All the above expressions contain $\dot{D}$ which is the material time derivative of the strain velocity tensor D. An effective use of the Eqs. (8)—(11) requires $\dot{D}$ to be expressed in terms of the velocity and the acceleration fields. To this end we will first evaluate $\dot{L}$, the material time derivative of L, and then will make use of Eq. (5a):

$$\dot{L} \equiv \frac{d}{dt}\left(v_{,k} \otimes q^k\right) = \left(\frac{d}{dt} v_{,k}\right) \otimes q^k + v_{,k} \otimes \frac{d}{dt} q^k = a_{,k} \otimes q^k + v_{,k} \otimes \dot{q}\,, \tag{12}$$

where a is the acceleration field. From the Eqs $(1\text{b})_3$ and $(5\text{b})_1$ we can deduce that

$$\dot{q}^k = -L^T q^k \tag{13}$$

and consequently

$$L = a_{,k} \otimes q^k - v_{,k} \otimes (L^T q^k)\,. \tag{14}$$

Since

$$v_{,k} \otimes (L^T q^k) = (v_{,k} \otimes q^k) L \equiv L^2\,, \tag{15}$$

we come up with the conclusion that

$$\dot{L} = \nabla a - L^2\,. \tag{16}$$

Symmetrization of the above equation (see Eq. (5a)) yields

$$\dot{D} = \tfrac{1}{2}\left[(\nabla a)^T + \nabla a - L^2 - (L^T)^2\right]. \tag{17}$$

The last result, when introduced to the Eqs. (8)—(11), permits the objective time derivatives to be expressed in terms of the velocity and acceleration gradients. The gradients can be effectively evaluated; if this is done in presence of appropriate kinematical constraints, the generalized strain accelerations, associated with the constraint, can be derived.

3. Strain acceleration measures in arbitrarily shaped shells

The tensors $\nabla \boldsymbol{a}, \boldsymbol{D}, \boldsymbol{W}, \boldsymbol{L}$, needed to obtain the objective measures of the strain acceleration tensor for shells, will be derived under the KIRCHHOFF assumption which — for thin shells — can also be employed in the large deformation range. The position vector of any particle of the shell is therefore described by the formula, Fig. 1,

$$\hat{\boldsymbol{r}}(X^k, t) = \boldsymbol{r}(X^\Delta, t) + \boldsymbol{d}(X^\Delta, t)\, X^3 \,, \tag{18a}$$

where

$$\boldsymbol{d} \cdot \boldsymbol{a}_\Sigma = 0 \,, \qquad \boldsymbol{a}_\Sigma \equiv \boldsymbol{r}_{,\Sigma} \,, \tag{18b}$$

$$\boldsymbol{d} \cdot \boldsymbol{d} = 1 \,. \tag{18c}$$

Beginning with the above equations capital, greek letters will take the values $\{1, 2\}$.

The velocity and the acceleration of a point of the shell are

$$\hat{\boldsymbol{v}}(X^k, t) \equiv \dot{\hat{\boldsymbol{r}}} = \dot{\boldsymbol{r}} + \dot{\boldsymbol{d}} X^3 \equiv \boldsymbol{v} + {}'\boldsymbol{v} X^3 = (v^\Delta \boldsymbol{a}_\Delta + v^3 \boldsymbol{d}) + ({}'v^\Delta \boldsymbol{a}_\Delta + {}'v^3 \boldsymbol{d})\, X^3 \,, \tag{19a}$$

$$\hat{\boldsymbol{a}}(X^k, t) \equiv \ddot{\hat{\boldsymbol{r}}} = \ddot{\boldsymbol{r}} + \ddot{\boldsymbol{d}} X^3 \equiv \boldsymbol{a} + {}'\boldsymbol{a} X^3 = (a^\Delta \boldsymbol{a}_\Delta + a^3 \boldsymbol{d}) + ({}'a^\Delta \boldsymbol{a}_\Delta + {}'a^3 \boldsymbol{d})\, X^3 \,. \tag{19b}$$

Their forms not containing ${}'v^\Delta$, ${}'v^3$ and ${}'a^\Delta$, ${}'a^3$, respectively, can bé obtained if use is made of the constraints (18 b, c). Differentiation of Eq. (18c) leads to the conclusion

$${}'v^3 = 0 \,, \tag{20}$$

whereas the time differentiation of the Eqs. (18b) gives

$${}'v_\Delta = \Omega_\Delta \,, \qquad \Omega_\Delta \equiv -v^3_{,\Delta} - B_{\Delta\Gamma} v^\Gamma \tag{21}$$

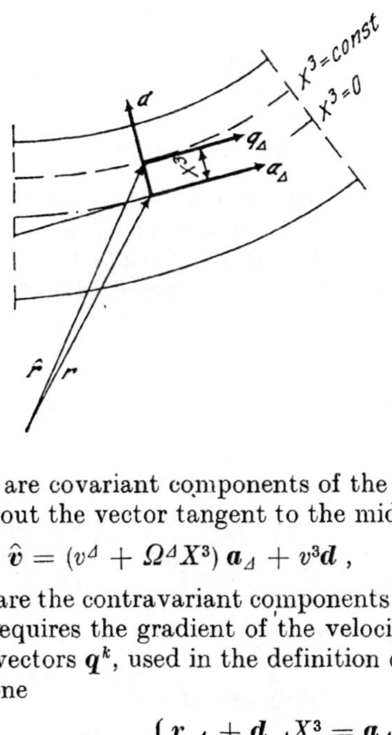

Fig. 1.

where $B_{\Delta\Gamma}$ are covariant components of the curvature tensor. The Ω_Δ, $\Delta = 1, 2$ represent the rotation of the shell element about the vector tangent to the middle surface of the shell. Finally

$$\hat{\boldsymbol{v}} = (v^\Delta + \Omega^\Delta X^3)\, \boldsymbol{a}_\Delta + v^3 \boldsymbol{d} \,, \qquad \Omega^\Delta \equiv \Omega_\Gamma a^{\Gamma\Delta} \,, \tag{22}$$

where $a^{\Delta\Delta}$ are the contravariant components of the metric tensor of the shell. A similar specification for the acceleration field requires the gradient of the velocity field to be computed first.

The vectors $\boldsymbol{q}^k$, used in the definition of the gradient of the velocity field, Eq. (6), form the basis dual to the following one

$$\boldsymbol{q}_k \equiv \hat{\boldsymbol{r}}_{;k} = \begin{cases} \boldsymbol{r}_{,\Delta} + \boldsymbol{d}_{,\Delta} X^3 = \boldsymbol{a}_\Delta - B_\Delta{}^\Sigma \boldsymbol{a}_\Sigma X^3 = (\delta_\Delta{}^\Sigma - B_\Delta{}^\Sigma X^3)\, \boldsymbol{a}_\Sigma \,, & k = \Delta \\ \boldsymbol{d}, \quad k = 3 \,. \end{cases} \tag{22}$$

Obviously $\boldsymbol{q}_\Delta = \boldsymbol{a}_\Delta$ for $X^3 = 0$. To specify the basis $\boldsymbol{q}^k$ compute first the metric tensor for the shell as a threedimenstional body

$$\begin{aligned} g_{\Delta\Gamma} &= [(\delta_\Delta{}^\Sigma - B_\Delta{}^\Sigma X^3)\, \boldsymbol{a}_\Sigma] \cdot [(\delta_\Gamma{}^A - B_\Gamma{}^A X^3)\, \boldsymbol{a}_A] \\ &= a_{\Sigma A}\, [\delta_\Delta{}^\Sigma \delta_\Gamma{}^A - (B_\Delta{}^\Sigma \delta_\Gamma{}^A + B_\Gamma{}^A \delta_\Delta{}^\Sigma)\, X^3 + B_\Gamma{}^A B_\Delta{}^\Sigma (X^3)^2] \,, \end{aligned} \tag{23a}$$

$$g_{\Delta 3} = g_{3\Delta} = 0 \,, \qquad g_{33} = 1 \,,$$

where

$$a_{\Sigma A} = \boldsymbol{a}_\Sigma \cdot \boldsymbol{a}_A = g_{\Sigma A}|_{X^3=0} \tag{23b}$$

are the components of the metric tensor of the middle surface of the shell. In what follows, the terms containing $(X^3)^2$ will be disregarded and the equalities obtained so will be denoted by $\simeq$. This means that the shell theory operating with the first moments only is admitted; for thin shells, considered herein, it is an appropriate assumption. Thus

$$g_{\Delta\Lambda} \simeq a_{\Delta\Gamma} - 2B_{\Delta\Gamma}X^3 . \tag{24}$$

The contravariant components $g^{\Delta\Lambda}$ form the matrix which is the inverse of that formed by $g_{\Delta\Lambda}$. It can be verified by inspection, that — within the accuracy assumed — the components $g^{\Delta\Lambda}$ are

$$g^{\Delta\Lambda} \simeq a^{\Delta\Lambda} + 2B^{\Delta\Gamma}X^3 , \qquad g^{\Delta 3} = g^{3\Delta} = 0 , \qquad g^{33} = 1 \tag{25}$$

with $a^{\Delta\Lambda}$ forming the inverse matrix with respect to that formed by $a_{\Delta\Lambda}$. Thus the dual basis is

$$\begin{aligned} \boldsymbol{q}^{\Delta} &= g^{\Delta\Gamma}\boldsymbol{q}_{\Gamma} \simeq (a^{\Delta\Gamma} + 2B^{\Delta\Gamma}X^3)\,(\delta_{\Gamma}{}^{\Sigma} - B_{\Gamma}{}^{\Sigma}X^3)\,\boldsymbol{a}_{\Sigma} \simeq (\delta_{\Lambda}{}^{\Delta} + B_{\Lambda}{}^{\Delta}X^3)\,\boldsymbol{a}^{\Lambda} , \\ \boldsymbol{q}^3 &= \boldsymbol{d} = \boldsymbol{q}_3 . \end{aligned} \tag{26}$$

According to Eq. (6) the gradient of the velocity field is

$$\begin{aligned} \boldsymbol{L} = \hat{\boldsymbol{v}}_{,k} \otimes \boldsymbol{q}^k &= \hat{\boldsymbol{v}}_{,\Lambda} \otimes \boldsymbol{q}^{\Lambda} + \hat{\boldsymbol{v}}_{,3} \otimes \boldsymbol{d} = [(v^{\Delta} + \Omega^{\Delta}X^3)\,\boldsymbol{a}_{\Delta} + v^3\boldsymbol{d}]_{,\Lambda} \otimes \boldsymbol{q}^{\Lambda} \\ &+ [(v^{\Delta} + \Omega^{\Delta}X^3)\,\boldsymbol{a}_{\Delta} + v^3\boldsymbol{d}]_{,3} \otimes \boldsymbol{d} . \end{aligned} \tag{27}$$

Evaluation of the above formula yields

$$\boldsymbol{L} \simeq [v_{\Delta}\|_{\Sigma} - B_{\Sigma\Delta}v^3 + X^3(\Omega_{\Delta}\|_{\Sigma} + B_{\Sigma}{}^{\Gamma}(v_{\Delta}\|_{\Gamma} - B_{\Gamma\Delta}v^3))]\,\boldsymbol{a}^{\Delta} \otimes \boldsymbol{a}^{\Sigma} - \Omega_{\Sigma}\boldsymbol{d} \otimes \boldsymbol{a}^{\Sigma} + \Omega_{\Sigma}\boldsymbol{a}^{\Sigma} \otimes \boldsymbol{d} , \tag{28}$$

where the symbol $\|$ stands for so called absolute derivative on the middle surface of the shell. For a vector field $\boldsymbol{s}$ the derivative is defined by the expression

$$s^{\Delta}\|_{\Lambda} = s^{\Delta}{}_{,\Lambda} + s^{\Sigma}\begin{Bmatrix} \Delta \\ \Sigma\Lambda \end{Bmatrix} , \tag{29}$$

where $\begin{Bmatrix} \Delta \\ \Sigma\Lambda \end{Bmatrix}$ are the Christoffel symbols.

Symmetric and antisymmetric part of the tensor $\boldsymbol{L}$ are

$$\begin{aligned} \boldsymbol{D} &\simeq ({}^{\varepsilon}D_{\Delta\Sigma} + X^{3\varkappa}D_{\Delta\Sigma})\,\boldsymbol{a}^{\Delta} \otimes \boldsymbol{a}^{\Sigma} , \\ \boldsymbol{W} &\simeq [\text{ant}\,(v_{\Delta}\|_{\Sigma}) + X^3(\text{ant}\,(\Omega_{\Delta}\|_{\Lambda}) + \text{ant}\,(B_{\Sigma}{}^{\Gamma}(v_{\Delta}\|_{\Gamma} - B_{\Gamma\Delta}v^3)))]\,\boldsymbol{a}^{\Delta} \otimes \boldsymbol{a}^{\Sigma} - \Omega_{\Sigma}\boldsymbol{d} \otimes \boldsymbol{a}^{\Sigma} + \Omega_{\Sigma}\boldsymbol{a}^{\Sigma} \otimes \boldsymbol{d} \end{aligned} \tag{30, 31}$$

with

$${}^{\varepsilon}D_{\Delta\Sigma} \equiv \text{sym}\,(v_{\Delta}\|_{\Sigma}) - B_{\Sigma\Delta}v^3 , \qquad {}^{\varkappa}D_{\Delta\Sigma} \equiv \text{sym}\,(\Omega_{\Delta}\|_{\Sigma}) + \text{sym}\,[B_{\Sigma}{}^{\Gamma}(v_{\Delta}\|_{\Gamma} - B_{\Gamma\Delta}v^3)] \tag{32, 33}$$

being the velocities of the membrane and bending strains respectively. The last two expressions are identical with those describing the same quantities in the linear shell theory. However, the curvature tensor $B_{\Sigma\Delta}$, entering the equations, is constant in the linear theory whereas it changes with time in the nonlinear analysis.

Having the velocity gradient $\boldsymbol{L}$ defined one is able to find the distribution of the acceleration field. To this end Eq. (18c) must be differentiated twice; the following equation is then obtained

$$\dot{\boldsymbol{d}} \cdot \boldsymbol{d} + \boldsymbol{d} \cdot \ddot{\boldsymbol{d}} = 0 \tag{34}$$

which, in view of Eqs. (19 a, b), results in

$${}'a^3 = \boldsymbol{d} \cdot \ddot{\boldsymbol{d}} = -\dot{\boldsymbol{d}} \cdot \dot{\boldsymbol{d}} = -{}'v^{\Delta}{}'v_{\Delta} = -\Omega^{\Delta}\Omega_{\Delta} . \tag{35}$$

Likewise, from the Eq. $(18\text{b})_1$ one can obtain

$$\ddot{\boldsymbol{d}} \cdot \boldsymbol{a}_{\Delta} + 2\dot{\boldsymbol{d}} \cdot \dot{\boldsymbol{a}}_{\Delta} + \boldsymbol{d} \cdot \ddot{\boldsymbol{a}}_{\Delta} = 0 . \tag{36}$$

If use is made of the property

$$\dot{\boldsymbol{a}}_{\Delta} = \boldsymbol{L}|_{X^3=0} \cdot \boldsymbol{a}_{\Delta} \Rightarrow \dot{\boldsymbol{a}}_{\Delta} = (v_{\Lambda}\|_{\Delta} - B_{\Lambda\Delta}v^3)\,\boldsymbol{a}^{\Lambda} - \Omega_{\Delta}\boldsymbol{d} \tag{37}$$

(which can be deduced from Eqs (1a) and (4)) and if the following relationships are taken into account

$$\ddot{\boldsymbol{d}} \cdot \boldsymbol{a}_{\Delta} = {}'a_{\Delta} , \qquad \boldsymbol{d} \cdot \ddot{\boldsymbol{a}}_{\Delta} = (\boldsymbol{d} \cdot \ddot{\boldsymbol{r}})_{,\Delta} - \boldsymbol{d}_{,\Delta} \cdot \ddot{\boldsymbol{r}} = \boldsymbol{a}^3_{,\Delta} + (B_{\Delta}{}^{\Gamma}\boldsymbol{a}_{\Gamma}) \cdot \boldsymbol{a} = a^3_{,\Delta} + B_{\Delta}{}^{\Gamma}a_{\Gamma} , \tag{38, 39}$$

$$\dot{\boldsymbol{d}} = {}'v^{\Gamma}\boldsymbol{a}_{\Gamma} = \Omega^{\Gamma}\boldsymbol{a}_{\Gamma} , \tag{40}$$

the expression defining the acceleration field is obtained in the form

$$\hat{\boldsymbol{a}} \simeq [a^{\Delta} - (2\Omega^{\Gamma}(v_{\Gamma}\|_{\Sigma}\,a^{\Sigma\Delta} - B_{\Gamma}{}^{\Delta}v^3) - \Gamma^{\Delta})\,X^3]\,\boldsymbol{a}_{\Delta} + (a^3 - \Omega^{\Gamma}\Omega_{\Gamma}X^3)\,\boldsymbol{d} , \tag{41}$$

where

$$\Gamma^{\Lambda} = (-a^3_{,\Delta} - B_{\Delta}{}^{\Gamma}a_{\Gamma})\,a^{\Delta\Lambda} . \tag{42}$$

The geometrical nonlinearity is reflected in Eq. (41) by the fact that the acceleration field $\hat{\boldsymbol{a}}$ is dependent on both the acceleration $\boldsymbol{a}$ and the velocity $\boldsymbol{v}$ of the middle surface of the shell.

The gradient of the acceleration field can now be found in the manner presented for the velocity field, i.e.

$$\nabla\hat{\boldsymbol{a}} = \hat{\boldsymbol{a}}_{,\Delta} \otimes \boldsymbol{q}^{\Delta} + \hat{\boldsymbol{a}}_{,3} \otimes \boldsymbol{d} . \tag{43}$$

Evaluation of the above formula along with its symmetrization leads eventually to the following expression

$$\begin{aligned}\operatorname{sym}(\nabla\hat{\boldsymbol{a}}) = \{&\operatorname{sym}(a_{\Delta}||_{\Lambda}) - B_{\Lambda\Delta}a^3 - X^3[2\operatorname{sym}[(\Omega^{\Pi}(v_{\Pi}||_{\Delta} - B_{\Pi\Delta}v^3))||_{\Lambda}] - \\ &- \operatorname{sym}(\Gamma_{\Delta}||_{\Lambda}) - B_{\Lambda\Delta}\Omega_{\Pi}\Omega^{\Pi} - \operatorname{sym}[B_{\Lambda}{}^{\Gamma}(a_{\Delta}||_{\Gamma} - B_{\Gamma\Delta}a^3)]]\}\,\boldsymbol{a}^{\Delta}\otimes\boldsymbol{a}^{\Lambda} + \\ &+ \tfrac{1}{2}\{-2\Omega^{\Pi}(v_{\Pi}||_{\Delta} - B_{\Pi\Delta}v^3) - X^3[(\Omega_{\Pi}\Omega^{\Pi})||_{\Delta} + 2\Omega^{\Pi}(v_{\Pi}||_{\Sigma} - B_{\Pi\Sigma}v^3)\,B_{\Delta}{}^{\Sigma}]\}\,\boldsymbol{d}\otimes\boldsymbol{a}^{\Delta} + \\ &+ \{\text{same as coef. of } \boldsymbol{d}\otimes\boldsymbol{a}^{\Delta}\}\,\boldsymbol{a}^{\Delta}\otimes\boldsymbol{d} - \Omega_{\Delta}\Omega^{\Delta}\boldsymbol{d}\otimes\boldsymbol{d}\,. \qquad (44)\end{aligned}$$

Once the tensors $\operatorname{sym}(\nabla\breve{\boldsymbol{a}})$, $\boldsymbol{L}$, $\boldsymbol{D}$ and $\boldsymbol{W}$ are known, any of the objective time derivatives discussed can be found according to one Eqs. (8)—(11) and Eq. (17). The analysis is already straightforward; the resulting expressions will therefore be demonstrated for the ZAREMBA-JAUMANN derivative only. The derivative is particulary well suited for rigid-ideally plastic rods and axisymmetric shells [10, 12].

According to Eqs. (8) and (17) the following relationships are obtained

$$\mathring{D} \cong ({}^{\varepsilon}\mathring{D}_{\Delta\Sigma} + X^3\,{}^{\varkappa}\mathring{D}_{\Delta\Sigma})\,\boldsymbol{a}^{\Delta}\otimes\boldsymbol{a}^{\Sigma} \qquad (45\mathrm{a})$$

with

$$\begin{aligned}{}^{\varepsilon}\mathring{D}_{\Delta\Sigma} = {}&\operatorname{sym}(a_{\Delta}||_{\Sigma}) - B_{\Delta\Sigma}a^3 + \Omega_{\Delta}\Omega_{\Sigma} + (v^{\Lambda}||_{\Delta} - B_{\Delta}{}^{\Lambda}v^3)(v_{\Lambda}||_{\Sigma} - B_{\Sigma\Lambda}v^3) - \\ &- 2[\operatorname{sym}(v_{\Lambda}||_{\Delta}) - B_{\Delta\Lambda}v^3][\operatorname{sym}(v^{\Lambda}||_{\Sigma}) - B^{\Lambda}{}_{\Sigma}v^3]\,, \qquad (45\mathrm{b})\end{aligned}$$

$$\begin{aligned}\mathring{D}_{\Delta\Sigma} = {}&\operatorname{sym}(\Gamma_{\Delta}||_{\Sigma}) + B_{\Delta\Sigma}\Omega_{\Pi}\Omega^{\Pi} - 2\operatorname{sym}[\Omega^{\Pi}(v_{\Pi}||_{\Delta}) - B_{\Pi\Delta}v^3)||_{\Sigma}] + \\ &+ \operatorname{sym}[B_{\Sigma}{}^{\Gamma}(a_{\Delta}||_{\Gamma}) - B_{\Gamma\Delta}a^3)] + B_{\Delta}{}^{\Gamma}(v^{\Lambda}||_{\Gamma} - B_{\Gamma}{}^{\Lambda}v^3)(v_{\Lambda}||_{\Sigma} - B_{\Sigma\Lambda}v^3) + \\ &+ B_{\Sigma}{}^{\Gamma}(v^{\Lambda}||_{\Delta} - B_{\Delta}{}^{\Lambda}v^3)(v_{\Lambda}||_{\Gamma} - B_{\Gamma\Lambda}v^3) - 4\operatorname{sym}\{[\operatorname{sym}(v_{\Lambda}||_{\Delta}) \\ &- B_{\Lambda\Delta}v^3][\operatorname{sym}(\Omega^{\Lambda}||_{\Sigma}) + \operatorname{sym}B_{\Sigma}{}^{\Gamma}(v^{\Lambda}||_{\Gamma} - B_{\Gamma}{}^{\Lambda}v^3)]\}\,. \qquad (45\mathrm{c})\end{aligned}$$

In the above ${}^{\varepsilon}\mathring{D}_{\Delta\Lambda}$, Δ, $\Lambda = 1, 2$ describe the acceleration of the membrane strains in the shell whereas ${}^{\varkappa}\mathring{D}_{\Delta\Lambda}$, Δ, $\Lambda = 1, 2$ describe the acceleration of the curvature changes; both are referred to the current configuration and are related to ZAREMBA-JAUMANN derivative only.

4. The strain accelerations in axisymmetric shells and plane rod structures

Significant simplifications can be introduced if rotationally symmetric deformation of an axisymmetric shell is considered. This is related to the fact that many of the quantities appearing in Eqs. (45b, c) vanish. So, parametrizing the meridian of the shell by its arc length one obtains, Fig. 2,

$$[a_{\Delta\Lambda}] = \begin{bmatrix} 1 & 0 \\ 0 & r^2 \end{bmatrix}, \qquad a^{\Delta\Lambda} = \begin{bmatrix} 1 & 0 \\ 0 & \dfrac{1}{r^2} \end{bmatrix}, \qquad (46)$$

$$[B_{\Delta\Lambda}] = \begin{bmatrix} \dfrac{1}{R_1} & 0 \\ 0 & r\sin\varphi \end{bmatrix}, \qquad [B_{\Delta}{}^{\Lambda}] = \begin{bmatrix} \dfrac{1}{R_1} & 0 \\ 0 & \dfrac{\sin\varphi}{r} \end{bmatrix}, \qquad (47)$$

$$\{\Omega_{\Delta}\} = \begin{Bmatrix} \Omega_1 \\ \Omega_2 \end{Bmatrix} = \begin{Bmatrix} -\dfrac{v^1}{R_1} - v^3_{,1} \\ 0 \end{Bmatrix}, \qquad \{\Gamma_{\Delta}\} = \begin{Bmatrix} \Gamma_1 \\ \Gamma_2 \end{Bmatrix} = \begin{Bmatrix} -\dfrac{a'}{R_1} - a^3_{,1} \\ 0 \end{Bmatrix}, \qquad (48)$$

where $(\;)_{,1}$ denotes the differentiation with respect to the arc. The covariant derivatives $v_{\Lambda}||_{\Delta}$ form the following matrix

$$[v_{\Lambda}||_{\Delta}] = \begin{bmatrix} v_{1,1} & 0 \\ 0 & v^1 r\cos\varphi \end{bmatrix}. \qquad (49)$$

Similar matrices, after appropriate change of symbols, are formed by $a_{\Lambda}||_{\Delta}$, $\Omega_{\Lambda}||_{\Delta}$ and $\Gamma_{\Lambda}||_{\Delta}$.

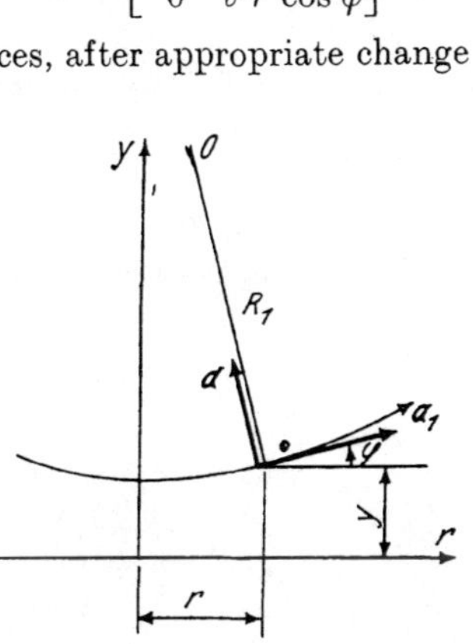

Fig. 2.

All the above relationships, when used in Eqs. (45 b, c), permit to find the nonzero components of the strain acceleration tensors, associated with the Zaremba-Jaumann derivative

$$ {}^{\varepsilon}\mathring{D}_{11} = a_{1,1} - \frac{a^3}{R_1} + \Omega_1^2 - \left(v_{1,1} - \frac{v^3}{R_1}\right)^2, \qquad {}^{\varepsilon}\mathring{D}_{22} = r a_r - (v_r)^2 , \qquad (50\text{a}), (50\text{b}) $$

$$ {}^{\varkappa}\mathring{D}_{11} = \Gamma_{1,1} + \frac{1}{R_1}\left[a_{1,1} - \frac{a^3}{R_1} - 2\left(v_{1,1} - \frac{v^3}{R_1}\right)^2 + \Omega_1^2\right] - 4\left(v_{1,1} - \frac{v^3}{R_1}\right)\Omega_{1,1} , \qquad (50\text{c}) $$

$$ {}^{\varkappa}\mathring{D}_{22} = \Gamma_1 r \cos\varphi + \Omega_1^2 r \sin\varphi + a_r \sin\varphi - 2\frac{\sin\varphi}{r} v_r^2 - 4\Omega_1 v_r \cos\varphi . \qquad (50\text{d}) $$

The quantities a_1, a_3 $(= a^1, a^3)$ as well as v_1, v_3 $(= v^1, v^3)$ are the components of the acceleration and the velocity vector tangent and perpendicular to the meridian respectively, while a_r, a_φ and v_r, v_φ are their components in radial and axial direction.

The kinematical relationships for plane rod structures coincide with those describing the meridional components for shells of revolution. In case of the Zaremba-Jaumann derivative they are given by Eqs (50a) and (50c).

5. Concluding remarks

A systematic method of deriving the objective strain acceleration measures for structurs, presented herein, differs from those usually employed in structural analysis. The measures are commonly derived by successive differentiation of components of the strain tensor. Although depending on the components, undergoing the differentiation, the objective rates can also be obtained, [5], such a method seems to be more involved. The complexity is associated with the fact the expressions like those of Eqs. (32) and (33), that then need to be differentiated, contain the time-dependent quantities $B_{\Delta\Lambda}$ and time-dependent operation $||$ (i.e. the covariant differentiation). Also, it is to be remembered that the time derivative of the component v_Δ of the velocity field is not a component of the acceleration field. So, although the idea of the successive differentiation of the components is clear, the above remarks indicate higher effectiveness of the method discussed in the paper.

It is obvious that the results like those of Eqs. (45 b, c) for the Zaremba-Jaumann derivative are too complicated to be used in practical computations. They have been cited only to fully present the method and they must be known to be able to simplify them further. The simplifications can be made by analysing either the order of the strains and rotations, [10], or the order of various components of the velocity and the acceleration vector (compare [2]).

References

1 Cotter, B. A., Rivlin, R. S., Tensors associated with time dependent stress, Quart. Appl. Math. **13** (1955) p. 177.
2 Duszek, M., Nonlinear equations of plastic shell theory (in Polish), IFTR Reports No 13, 1971.
3 Eringen, A. C. (ed.), Continuum physics, vol. **1** – Mathematics, Academic Press, New York-London, 1971.
4 Jaumann, G., Geschlossenes System physikalischer and chemischer Differentialgesetze, Sitzungsber. Akad. Wiss. Wien (20) **120** (1911) p. 385.
5 Masur, E. F., On the definition of stress rate, Quart. Appl. Math. **19** (1961) p. 160.
6 Noll, W.; Truesdell, C., The nonlinear field theory, Encyclopedia of Physics, Ed. S. Flügge, vol. III/3, Springer, Berlin-Göttingen-Heidelberg, 1965.
7 Oldroyd, J. G., On the formulation of the rheological equations of state, Proc. Roy. Soc. (A) **200** (1950) p. 523.
8 Prager, W., An elementary discussion of definitions of stress rate, Quart. Appl. Math. **18** (1961) p. 403.
9 Rivlin, R. S.; Ericksen, C., Stress-deformation relations for isotropic materials, J. Rat. Mech. Anal. **2**, **4** (1955) p. 323.
10 Stolarski, H.; Belytschko, T., Large deformation, rigid-plastic dynamics by an extremum principle, Comp. Meth. Appl. Mech. Eng. **2**, **21** (1980) p. 217.
11 Stolarski, H., An extremum principle in nonlinear dynamics of hardening rigid-plastic bodies, Proceedings of the IUTAM Symposium on Variational Methods in the Mechanics of Solids, Northwestern University, September 1978, p. 98.
12 Stolarski, H., An extremum principle for displacement estimation in dynamically loaded, rigid-plastic structures, (in Polish), IFTR Reports, 22, 1979.
13 Truesdell, C., Elasticity and fluid dynamics, J. Rat. Mech. Analysis **2** (1953) p. 593.
14 Zaremba, S., Le principe des mouvements relatifs et les équations de la mécanique physique, Bull. Int. Acad. Sci. Cracovie 1903, p. 614.

Eingereicht am 13. 1. 1981

Anschrift: Dr. H. Stolarski, Inst. of Fundamental Technological Research, Polish Academy of Sciences, Ul. Swietokrzyska 21 00-049 Warszawa, Poland

KLEINE MITTEILUNGEN

ZAMM 61, 657 – 659 (1981)

J. Bergmann, J. Socolowsky

Über den Druckverlauf im Breitschlitzextruder

1. Einleitung

Eine ganze Reihe technologischer Prozesse, so z. B. die Folieherstellung, der Filmbeguß u. a., benutzen als wesentliches Element das Breitschlitzextrusionswerkzeug (Abb. 1). Eine seiner wesentlichen Aufgaben besteht darin, das in der Mitte ($z = 0$) dem Werkzeug zugeführte fluide Medium über die gesamte Breite ($-b \leqq z \leqq b$) möglichst gleichmäßig zu verteilen. In dieser Arbeit beschäftigen wir uns mit dem Druckverlauf im Querströmrohr beim Fließen nichtnewtonscher Medien. Unter der Annahme, daß in jedem der beiden charakteristischen Teile dieses Werkzeugs (Querströmrohr, Austrittsspalt) die Strömung als stationär, inkompressibel, isotherm und eindimensional angenommen wird, wurde in [1] die folgende Druckgleichung

$$-\frac{dQ_R}{dz} = Q \tag{1}$$

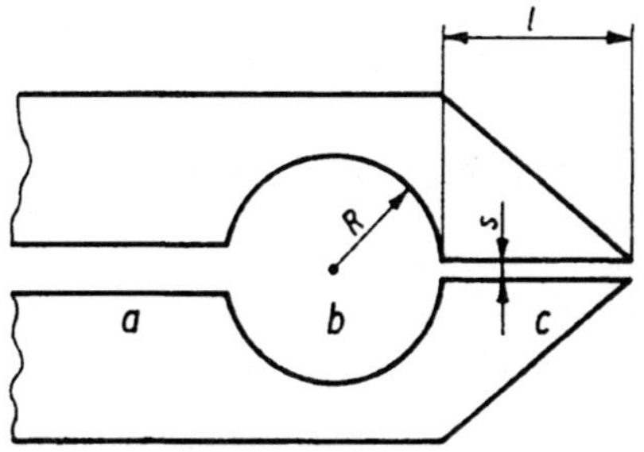

Abb. 1. Extruderquerschnitt bei $z = 0$
a — Einströmrohr, b — Querströmrohr, c — Austrittsspalt

hergeleitet und für Spezialfälle numerisch ausgewertet. In (1) bedeuten Q_R den Durchsatz im Rohr,

$$Q_R = Q_R(p') = \pi R^3 \int_0^1 \xi^2 f\left(-\frac{R}{2}\, p'(z)\, \xi\right) d\xi\,, \tag{2}$$

Q den Durchsatz im Spalt,

$$Q = Q(s, p) = 2^{-1} s^2 \int_0^1 \xi f\left(\frac{s}{2l}\, p(z)\, \xi\right) d\xi\,, \tag{3}$$

$p = p(z)$ den im Rohr herrschenden Überdruck und $f(t)$ die Materialfunktion. Die anderen auftretenden Größen sind in Abb. 1 erklärt, wobei wir annehmen werden, daß R, l Konstanten und $s = s(z)$ beschränkt und stückweise stetig sind. Bezüglich $w = f(t)$ sei vorausgesetzt:

V1. $f(t)$ — stetige, ungerade, streng monoton wachsende Funktion;

V2. $\exists \gamma > 0,\ C > 0$: $\lim\limits_{t\to 0} f(t)\, t^{-\gamma} = C$, $\lim\limits_{t\to\infty} f(t) = \infty$. Damit werden neben dem newtonschen Medium ($f(t) = t/\eta$) vor allem Potenzgesetzflüssigkeiten $\left(f(t) = \frac{t}{\eta^*}\left|\frac{t}{\eta^*}\right|^{1/n\,-1};\ \gamma = \frac{1}{n}\right)$, aber auch das nichtnewtonsche Verhalten nach Meter $\left(f(t) = \frac{t}{\eta_0}\left(1 + \left|\frac{t}{t_M}\right|^{\alpha-1}\right)\left(1 + \frac{\eta_\infty}{\eta_0}\left|\frac{t}{t_M}\right|^{\alpha-1}\right)^{-1};\ \gamma = 1\right)$ und Ellis $\left(f(t) = \frac{t}{\eta_0}\left(1 + \left|\frac{t}{t^*}\right|^{\alpha-1}\right);\ \gamma = \min(\alpha, 1)\right)$, sowie solche implizit gegebenen Materialfunktionen, wie z. B. vom Sutterbyschen Typ ($f^{-1}(w) = \eta_0 w \left[\frac{\operatorname{ar\,sinh}(Bw)}{Bw}\right]^A;\ \gamma = 1$) erfaßt. Viskoplastische Medien hingegen erfüllen V1., V2. nicht. Bezüglich der Materialfunktionen vgl. [2].

Aus der Beziehung, daß der Durchsatz im Querströmrohr bei $z = b$ verschwindet und bei $z = 0$ die Hälfte des zugeführten Durchsatzes D ist, ergeben sich als Randbedingungen

$$Q_R(0) = D/2 > 0\,, \qquad Q_R(b) = 0\,. \tag{4.1), (4.2}$$

Wegen V1. folgt aus (4.2) $p'(b) = 0$. Statt (4.1) wird manchmal auch

$$p(0) = p_0 > 0 \tag{4.1'}$$

vorgegeben [3]. Ziel der Arbeit ist es, Existenz- und Optimalitätsaussagen für die Lösung der Aufgabe (1)–(4) zu gewinnen.

2. Existenz der Lösung

Setzen wir

$$q(z) = -Q_R(p')\,, \tag{5}$$

so läßt sich (1), (4) als System aufschreiben,

$$\frac{dq}{dz} = Q(s(z), p(z)) \qquad \frac{dp}{dz} = P(q(z))\,, \tag{6.1), (6.2}$$

mit den Randbedingungen

$$q(0) = -D/2\,, \qquad q(b) = 0 \tag{7.1), (7.2}$$

bzw. (4.1′) statt (7.1). Die Gleichung (6.2) ergibt sich durch Auflösen von (5) nach p'. $P(t)$ und $Q(s, t)$ sind dabei bezüglich t stetige, ungerade, monoton wachsende Funktionen.

Definition: Unter der Lösung $p(z)$ der Aufgabe (1)–(4) verstehen wir eine stetig differenzierbare Funktion, welche die Randbedingungen (4.1), (4.2) bzw. (4.1′), (4.2) erfüllt, für die $Q_R(p')$ absolut stetig ist und welche fast überall (1) erfüllt. Diese Definition ist äquivalent dazu, daß p, q absolut stetig sind und ast überall (6.1), (6.2) erfüllen.

Satz 1: *Die Aufgabe (1)–(4) besitzt eine und nur eine Lösung.*

Für den Beweis von Satz 1 benötigen wir einige a-priori Aussagen über das Monotonieverhalten der Lösung $p(z)$ von (1)–(4).

Lemma 1: *$p(z)$ ist eine positive, streng monoton fallende Funktion.*

Beweis: Wir zeigen zunächst, daß aus $p(z_0) = 0$ oder $q(z_0) = 0$ sofort $p(z) \equiv q(z) \equiv 0$, $z \in [z_0, b]$ folgt. Dazu multiplizieren wir (6.1) mit $p(z)$ und integrieren von z_0 bis b:

$$p \cdot q\Big|_{z_0}^{b} = \int_{z_0}^{b} q(z)\, P(q(z))\, dz + \int_{z_0}^{b} p(z)\, Q(s, p(z))\, dz\,.$$

Da die linke Seite verschwindet, die Integranden auf der rechten Seite hingegen nichtnegativ sind, muß gelten $p(z) \equiv q(z) \equiv 0$, $z \in [z_0, b]$. Daraus folgt, daß beide Funktionen vorzeichenkonstant sind. Ist die Randbedingung (7.1) vorgegeben, so ist also $q(z) \leqq 0$ und mithin, wegen (6.2), $p'(z) \leqq 0$. Wäre jetzt $p(z) \leqq 0$, $z \in [0, b]$, so folgte aus (6.1) sofort $q'(z) \leqq 0$, was (7.1) in Widerspruch zu (7.2) setzt, also $p(z) \geqq 0$. Im Falle der Randbedingung (4.1′) ist die Nichtnegativität von $p(z)$ durch $p_0 > 0$ gesichert. Mit (6.1) wird dann aber $q'(z) \geqq 0$ fast überall, woraus wegen (7.2), (6.2) $q(z) \leqq 0$, $p'(z) \leqq 0$ folgen.

Um nun die in der Formulierung von Lemma 1 behaupteten strengen Ungleichungen zu beweisen, nehmen wir an, es gäbe einen Punkt $z_0 \in]0, b]$ mit der Eigenschaft, daß die Lösung der Aufgabe (6), (7) dort verschwindet, für alle $z < z_0$ hingegen $p(z) > 0$, $q(z) < 0$ gelten. Dann ist die Lösung der Randwertaufgabe (6), (7) auch Lösung der Anfangswertaufgabe (6.1), (6.2),

$$q(z_0) = 0\,, \qquad p(z_0) = 0\,. \tag{8.1), (8.2}$$

Eine Lösung der Aufgabe (6), (8) ist $p(z) \equiv q(z) \equiv 0$. In Lemma 2 zeigen wir, daß dies die einzige ist. Wir erhalten einen Widerspruch zu $p(z) > 0$ links von z_0. Daraus folgt, daß $p(z) \neq 0$ für alle $z \in [0, b]$ und mithin $q'(z) > 0$, woraus sich sofort $q(z) < 0$ und somit auch $p'(z) < 0$, $z \in [0, b[$ ergibt.

Lemma 2: *Die Lösung der Anfangswertaufgabe (6), (8) ist eindeutig.*

Beweis: Substituiert man $y = z_0 - z$ und bezeichnet alle Funktionen $f(z_0 - y)$ durch $f_1(y)$, so kann man (6), (8) in der

Form

$$\frac{dq_1}{dy} = -Q(s_1(y), p_1(y)), \quad q_1(0) = 0, \tag{9.1}$$

$$\frac{dp_1}{dy} = -P(q_1(y)), \quad p_1(0) = 0 \tag{9.2}$$

aufschreiben. Genau wie in Lemma 1 zeigt man, daß $p_1(y)$ und $q_1(y)$ für $y > 0$ vorzeichenkonstante Funktionen sind. O. B. d. A. gilt $p_1(y) \geqq 0$, wenn $y \geqq 0$. Wird (9.1) in (9.2) eingesetzt, ergibt sich

$$p_1(y) = \int_0^y P\left[\int_0^t Q(s_1(x), p_1(x))\, dx\right] dt.$$

Berücksichtigt man die Beschränktheit der Funktion $s(z)$,

$$0 < s_0 \leqq s(z) \leqq s^0, \quad s_0, s^0 = \text{const.}, \tag{10}$$

so ergibt sich für Q die Abschätzung $Q(s, p_1) \leqq Q^0(p_1) = Q(s^0, p_1)$, und wegen der Monotonie von P ist für jedes $c_1 > 0$

$$p_1(y) \leqq c_1 + \int_0^y P\left[\int_0^t Q^0(p_1(x))\, dx\right] dt = u(y);$$

$u(y)$ ist monoton wachsend in y und

$$\pi R^3 \frac{d}{dy} \int_0^1 \xi^2 f\left(\frac{R}{2} u' \xi\right) d\xi \leqq \frac{1}{2} (s^0)^2 \int_0^1 \xi f\left(\xi \frac{s^0}{2l} u\right) d\xi;$$

integriert man beide Seiten dieser Ungleichung von 0 bis u und substituiert im linken Integral $u'(y) = v(u)$, so wird

$$\pi R^3 \left[\int_0^1 \xi^2 f\left(\frac{R}{2} v(u)\, \xi\right) d\xi - \int_0^u \left(\int_0^1 \xi^2 f\left(\frac{R}{2} \tilde{u} v(\tilde{u})\, \xi\right) d\xi \frac{dv}{du}\right) d\tilde{u}\right] \leqq$$

$$\leqq \int_0^u \frac{1}{2} (s^0)^2 \int_0^1 \xi f\left(\xi \frac{s^0}{2l} \tilde{u}\right) d\xi\, d\tilde{u}.$$

Wegen V2. wird $|f(t)\, t^{-\gamma} - C| < \varepsilon$, wenn $0 < t < \delta$, d. h. $c_0 t^\gamma < f(t) < c^0 t^\gamma$. Dabei ist ε derart, daß $0 < c^0 < (1 + \gamma)\, c_0$. Wird nun u so klein gewählt, daß sowohl $s^0 u/l < \delta$ als auch $\frac{R}{2} v(u) < \delta$, so ist wegen $\frac{dv}{du} \geqq 0$

$$\pi R^3 \int_0^1 \xi^{2+\gamma} \left(\frac{R}{2}\right)^\gamma d\xi \left[c_0 - \frac{c^0}{1+\gamma}\right] v^{1+\gamma}(u) \leqq \frac{1}{2} c^0 (s^0)^2 \left(\frac{s^0}{2l}\right)^\gamma \times$$

$$\times \frac{u^{1+\gamma}}{1+\gamma} \int_0^1 \xi^{1+\gamma}\, d\xi$$

und somit $u'(y) \leqq K u(y)$ mit $K > 0$, woraus $u \leqq c_1 \exp(Ky)$ folgt. $c_1 \to 0$ liefert dann das Gewünschte.

Zum Beweis von Satz 1 benötigen wir schließlich noch ein Resultat von R. Conti [4], welches wir, bezogen auf die Aufgabe (6), (7) so formulieren können:

Die Lösung der Aufgabe

$$x' = A_0 x + g(z, x), \quad x \in R_2, \quad z \in [0, b],$$

$$A_1 x(0) + A_2 x(b) = \varphi$$

($A_i - i = 0, 1, 2 - 2 \times 2$ Matrizen, $g\colon [0, b] \times R_2 \to R_2$ erfüllt Caratheodorybedingungen) existiert, wenn

- $|g(z, x)| \leqq r(z) \in L(0, b)$,
- die Lösung der zugehörigen homogenen Aufgabe (d. h. $g = 0$, $\varphi = 0$) eindeutig ist.

Beweis von Satz 1: Das Gleichungssystem (6) schreiben wir in der Form

$$\frac{dq}{dz} = -p + p + Q(s, p),$$

$$\frac{dp}{dz} = P(q).$$

Identifiziert man $x = \begin{pmatrix} q \\ p \end{pmatrix}$, $g(z, x) = \begin{pmatrix} p + Q(s, p) \\ P(q) \end{pmatrix}$, so erkennt man sofort die Eindeutigkeit der Lösung der zugehörigen homogenen Aufgabe. Die gleichmäßige Beschränktheit von $g(z, x)$ kann erhalten werden, wenn es gelingt, a-priori die Beschränktheit der Lösung der Aufgabe (6), (8) nachzuweisen. Im Falle der Randbedingungen (4.1'), (7.2) folgt diese für $p(z)$ sofort aus Lemma 1: $0 < p(z) \leqq p_0$ und für $q(z)$ aus (6.1), (7.2): $-Q^0(p_0)\, b < q(z) \leqq 0$. Für die Randbedingungen (7.1), (7.2) ergibt Lemma 1 in Zusammenhang mit (6.1) nur eine Aussage für $q(z)$: $-D/2 \leqq q(z) \leqq 0$, $q(z)$ monoton steigend. Folglich ist $-P(D/2) \leqq p'(z) \leqq 0$ und $p(z) \geqq p_0 + P(-D/2)\, z$. Dann ist entweder $p_0 \leqq P(D/2)\, b$ und mithin $p(z)$ beschränkt, oder aus der Integration von (6.1),

$$\frac{D}{2} = \int_0^b Q(s(z), p(z))\, dz \geqq \int_0^b Q(s(z), p_0 + P(-D/2)\, z)\, dz,$$

erhält man wegen der Monotonie von $Q(s, t)$ eine obere Grenze für p_0 und somit die Beschränktheit von $p(z)$.

Die Eindeutigkeit der Lösung folgt aus der Monotonie der rechten Seiten von (6) (vgl. Satz 3 aus [5] und Anm. 1 dazu).

Anmerkung: Für die angegebenen konkreten Materialfunktionen lassen sich Existenz- und Eindeutigkeitsaussagen auch über Satz 2.2 aus [6] gewinnen.

3. Optimale Spaltgeometrie

Definition: Die Spaltgeometrie heißt druckoptimal, wenn

$$J(p) = \sup_{[0,b]} |p_0 - p(z)| \to \min_{s(z)}!,$$

sie heißt schichtdickenoptimal, wenn die extrudierte Schichtdicke konstant ist. Dabei ist $s(z)$ aus der Klasse der stückweise stetigen Funktionen, für die (10) gilt.

Bezüglich schichtdickenoptimaler Spaltgeometrie, welche u. a. bei der Film- oder Folieherstellung bedeutsam ist, siehe [7].

Satz 2: *Charakterisiere $s^*(z)$ die druckoptimale Spaltgeometrie, so ist $s^*(z) = s_0$.*

Beweis: Wegen Lemma 1 ist die vorgelegte Optimierungsaufgabe äquivalent einer Steueraufgabe vom Mayerschen Typ:

$$-p(b) \to \min_{s(z)}!, \quad q' = Q(s, p), \quad p' = P(q)$$

mit den entsprechenden Randbedingungen. Ist nun $s^*(z)$ die „optimale Steuerung", so maximiert sie die Hamiltonfunktion $H = \psi_1 Q + \psi_2 P$,

$$\frac{d\psi_1}{dz} = -\frac{dP}{dq}, \quad \psi_2(0) = 0, \tag{11.1}$$

$$\frac{d\psi_2}{dz} = -\frac{\partial Q}{\partial p}, \quad \psi_2(b) = 1. \tag{11.2}$$

Im Falle der Randbedingung (4.1') ist $\psi_2(0) = 0$ durch $\psi_1(0) = 0$ zu ersetzen. Analog zu Lemma 1 kann man auch für diese Randwertaufgabe eine Vorzeichenanalyse durchführen, da wegen der Monotonieeigenschaften der Funktionen P und Q die in (11) auftretenden Ableitungen dieser Funktionen für $z < b$ positiv sind. Insbesondere erhält man $\psi_1 < 0$ für $z > 0$. Damit wird aber das Maximum der Hamiltonfunktion für $s^*(z) = s_0$ angenommen.

Anmerkung: Bezieht man die Konstante R in die Optimierung mit ein, so läßt Satz 2 die Aussage zu, daß für alle nichtnewtonschen Medien, deren Materialfunktionen die Eigenschaften V1, V2 haben, die druckoptimale Geometrie des Werkzeugs durch maximales R/s charakterisiert werden kann. Entsprechend tritt der größte Druckverlust bei $s(z) = s^0$ auf. Damit ergibt sich die Möglichkeit, den Druckverlust J bei variabler Spaltbreite durch den Druckverlust bei konstanter Spaltbreite abzuschätzen. Für Potenzgesetzflüssigkeiten untersuchen wir den letzten Fall etwas genauer.

4. Potenzgesetzmedien

Für konstante (z-unabhängige) Extrudergeometrie lautet die Druckgleichung (1) unter Berücksichtigung der Aussagen von Lemma 1

$$p''(-p')^{m-1} = A_n p^m, \quad p'(b) = 0, \tag{12}$$

mit

$$A_n = \frac{s^{2+m}(1 + 3n)n}{l^m R^{3+m}(1 + 2n)\, 2\pi}, \quad m = \frac{1}{n}, \; B_n = A_n^{1/(1+m)}$$

Satz 3: *Die Lösung der Aufgabe* (12) *ist darstellbar als*

$$\sum_{k=l}^{\infty} \frac{a(z)^{km+k-1}}{km+k-1} = B_n(b-z) \tag{13}$$

mit $a(z) = (1 - (p(b)/p(z))^{1+m})^{1/6(1+m)}$.

Beweis: Unter Berücksichtigung der Randbedingung ergibt die einmalige Integration von (12)

$$p'(z) = -B_n(p(z)^{1+m} - p(b)^{1+m})^{1/(1+m)} . \tag{14}$$

Bezeichnet man

$$\alpha = p(z)/p(b) ,$$

so erhält man die Integraldarstellung

$$\int_{\alpha}^{1} (\xi^{1+m} - 1)^{-1/(1+m)} \, d\xi = B_n(z-b) ;$$

j-malige partielle Integration liefert

$$\sum_{k=1}^{j} \frac{a(z)^{km+k-1}}{km+k-1} - \int_{1}^{\alpha} (\xi^{m+1} - 1)^{(jm+j-1)/(1+m)} \, \xi^{-jm-j+1} \, d\xi$$

$$= B_n(b-z) .$$

Wegen Lemma 1 gilt für das im letzten Ausdruck auftretende Integral I die gleichmäßige Abschätzung bei $j \to \infty$

$$|I| \leqq (\alpha(0)^{1+m} - 1)^{(jm+j-1)/(1+m)} \, \alpha(0) \to 0 ,$$

woraus (13) folgt.

Formel (13) kann zur näherungsweisen Bestimmung des Druckes benutzt werden. Dazu setzt man in (13) $z = 0$ und kann dann iterativ $a(0) = a_0$ bestimmen. Ist die Randbedingung (4.1′) gegeben, so läßt sich aus a_0 sofort $p(b)$ berechnen. Im Falle (4.1) ergibt sich aus (4.1) und (14) die Beziehung

$$\left(\frac{Ra_0 p_0}{2\eta^*}\right)^m A_n^{1/(1+n)} \frac{R^3 n\pi}{1+3n} = \frac{D}{2} , \tag{15}$$

welche es gestattet, zunächst p_0 und damit auch $p(b)$ zu berechnen. Damit wird (13) zur Bestimmung von $p(z)$ anwendbar. Formel (15) gestattet es, die beiden betrachteten Randwertaufgaben ineinander überzuführen.

Für die Qualität der extrudierten Schicht ist ihre gleichmäßige Dicke von Bedeutung. Unter der Voraussetzung, daß diese direkt proportional dem Durchsatz im Austrittsspalt ist, läßt sich als Qualitätsmaß die Größe $I_n = Q(s, p(b))/Q(p(0)) < 1$ angeben [3]. Aus Satz 3 ergibt sich die

Folgerung: $\lim\limits_{n\to 0} I_n = 0$; $\lim\limits_{n\to\infty} I_n = 1$, wenn $3s^2 b < 4\pi R^3$.

Tatsächlich, aus (13) erhält man die Abschätzung

$$[1 - \exp(-m(b-z) B_n)]^n \leqq a(z) \leqq$$

$$\leqq [1 - \exp(-(m+1)(b-z) B_n)]^{1/(m+1)} ,$$

woraus sofort

$$\{1 - [1 - \exp(-(b-z) B_n/n)]^{1+n}\}^{1/(1+n)} \geqq I_n(z) \geqq$$

$$\exp(-(b-z) B_n/n),$$

$(I_n(z) = Q(s, p(b))/Q(s, p(z)))$ und damit die gleichmäßige Konvergenz $I_n(z) \to 0$ für $z \in [0, b-\varepsilon]$, $\varepsilon > 0$ folgt.

Berücksichtigt man in (13) nur das erste Glied der Summe, so ergibt sich

$$1 > I_n > [1 - (B_n b/n)^{1+n}]^{1/(1+n)} .$$

Bei $n \to \infty$ wird $B_n b/n < 1$, woraus sich $\lim\limits_{n\to\infty} I_n = 1$ ableitet.

Damit kann die Schlußfolgerung gezogen werden, daß Potenzgesetzflüssigkeiten mit hohem Fließexponenten vermittels des Breitschlitzextruders eine gleichmäßigere Verteilung erfahren (die in der Folgerung geforderte Beschränkung bez. der Werkzeugparameter ist in der Regel erfüllt). Für $n = 1$ folgt aus (13) durch einfache Umformungen die bereits in [3] für I_1 erhaltene Formel $I_1 = \cosh^{-1} \sqrt{A_n}\, b$.

Literatur

1 BERGMANN, J.; LIPP, R.; REHER, E.-O., Über Toleranzmaße beim Filmbeguß I, Journal f. Signalaufzeichnungsmaterialien, in Vorbereitung.

2 REHER, E.-O.; KÄRMER, R., Mechanische Verfahrenstechnik, 1. Lehrbrief, hrsg. v. d. Zentralst. f. d. Hochschulfernstudium, Dresden 1975.

3 LIPP, R.; REHER, E.-O., Zur Strömung von Herschel-Bulkley-Medien in mehrstufigen Extrusionswerkzeugen, Plaste und Kautschuk **26** (1979) 7, S. 391–393.

4 CONTI, R., Equazioni differenziali ordinarie quasilineari con condizioni lineari, Ann. mat. pura ed appl. **57** (1962), S. 49–61.

5 *Васильев, Н. И.; Клоков, Ю. А.* Основы теории краевых задач обыкновенных дифференциальных уравнений, глава 6, Зинатне, Рига 1978.

6 GAJEWSKI, H.; GRÖGER, K.; ZACHARIAS, K., Nichtlineare Operatorgleichungen und Operatordifferentialgleichungen, Kapitel III, Akademie-Vlg., Berlin 1974.

7 BERGMANN, J., LIPP, R., SCHULZE, J., Über Toleranzmaße beim Filmbeguß II, in Vorbereitung.

Eingereicht am 28. 10. 1980

Anschriften: Dr. J. BERGMANN, Dipl.-Math. J. SOCOLOWSKY, Technische Hochschule „Carl Schorlemmer“ Leuna-Merseburg, Sektion Mathematik, 4200 Merseburg, Geusaer Str., DDR

ZAMM 61, 659–660 (1981)

S. AIZICOVICI / C. FETECĂU

Existence of Couette Flows in Incompressible Simple Fluids

The aim of this note is to provide an existence theory for a class of Couette flows in incompressible simple fluids of integral type, the uniqueness of which had previously been established by one of the authors (cf. [1], [2]).

1. We begin by briefly indicating the types of flows we will be concerned with in the sequel. The constitutive equation of an incompressible simple fluid of integral type of the first order is ([3], § 37)

$$\boldsymbol{T} = -p\boldsymbol{1} + \int_0^{\infty} \zeta(s) \, \boldsymbol{G}(s) \, ds , \tag{1}$$

where $\boldsymbol{T}$ is the stress tensor, p is an indeterminate pressure, $\boldsymbol{G}(\cdot) = \boldsymbol{F}_t^t(\cdot)^T \boldsymbol{F}_t^t(\cdot) - \boldsymbol{1}$, with $\boldsymbol{F}_t^t(\cdot)$ the relative deformation gradient history and $\boldsymbol{1}$ the unit tensor while $\zeta(\cdot) \leqq 0$ is a material function. The flows to be considered here have the velocity fields ([1], [2])

$$\dot{x} = 0 , \qquad \dot{y} = u(x, t) , \qquad \dot{z} = v(x, t) \tag{2}$$

and

$$\dot{r} = 0 , \qquad \dot{\theta} = \omega(r, t) , \qquad \dot{z} = w(r, t) , \tag{3}$$

in Cartesian and cylindrical co-ordinates, respectively.

First, we confine our attention to a bilineal flow between two infinite parallel plates, which till $t = 0$ had been in rest and after this moment they had moved with assigned speeds $U(t)$ and $V(t)$ in the directions y and z, respectively. Assuming that the fluid adheres to the plates one can prove (see [1]) that the flow is characterized by the integral equation

$$\int_0^t \eta(t-\sigma) \, \partial_x^2 \varphi(x, \sigma) \, d\sigma + f(x, t) = \partial_t \varphi(x, t) , \qquad x \in [0, d] ,$$
$$t \geqq 0 , \tag{4}$$

together with the initial and boundary conditions

$$\left.\begin{aligned} &\varphi(x, 0) = 0 , && x \in [0, d]; \\ &\varphi(0, t) = \varphi(d, t) = 0 , && t \geqq 0 . \end{aligned}\right\} \tag{5}$$

Here $\eta(t) = -\dfrac{1}{\varrho} \displaystyle\int_t^{\infty} \zeta(s) \, ds$, with ϱ the mass density, $\varphi(x, t)$ is any of $u(x, t) - \dfrac{x}{d} U(t)$ and $v(x, t) + \dfrac{x-d}{d} V(t)$, where d is the distance between plates and $f(x, t)$ is in each case a known function.

Consider next a flow between two infinite cylinders, which at $t = 0$ had begun to rotate about their common axis ($r = 0$) and slide in the direction of the same axis with assigned velocities $\Omega_1(t)$, $\Omega_2(t)$ and $W_1(t)$, $W_2(t)$, respectively. Assuming again that the fluid adheres to the walls we are led (cf. [2]) to the initial-value problem

$$\int_0^t \eta(t-s)\, L_n\varphi_n(r,s)\,ds + f_n(r,t) = \partial_t\varphi_n(r,t)\,, \qquad r \in [R_1, R_2]\,, \quad t \geqq 0\,, \tag{6}$$

$$\left.\begin{aligned} &\varphi_n(r,0) = 0\,, \quad r \in [R_1,R_2]\,,\\ &\varphi_n(R_1,t) = \varphi_n(R_2,t) = 0\,, \quad t \geqq 0\,, \end{aligned}\right\} \tag{7}$$

where $n = 0, 1$, $\varphi_0(r,t) = w(r,t) + \frac{1}{R_2 - R_1}[(r - R_2)\, W_1(t) - (r - R_1)\, W_2(t)]$, $\varphi_1(r,t) = r\omega(r,t) + \frac{r}{R_2 - R_1}[(r - R_2) \times \Omega_1(t) - (r - R_1)\,\Omega_2(t)]$, $L_n = \partial_r^2 + \frac{1}{r}\partial_r - \frac{n}{r^2}$, $f_n(r,t)$ are also known functions, R_1 and R_2 denote the cylinders radii, while η is the same as above.

2. We are going to prove the existence of solutions to problems (4), (5) and (6), (7). The key idea is to view the equations (4) and (6) as abstract Volterra integrodifferential equations of the form

$$\frac{du}{dt} + \int_0^t \eta(t-s)\, Au(s)\,ds = f(t)\,, \qquad t \geqq 0; \qquad u(0) = 0\,, \tag{8}$$

where u and f take values in a Hilbert space H, A is a linear, unbounded operator in H and the integral is interpreted in the Bochner sense. The boundary conditions (5_2) and (7_2) are to be incorporated in the domain of A. We assume the reader is familiarized with the concepts of vector-valued function, Bochner integral, Sobolev space, as well as with the usual functional notations we use below. For details, see, e.g. [4], Ch. 1.

Let V and H be real Hilbert spaces such that $V \subset H$, with dense and continuous inclusion. Identifying H with its own dual we may write $V \subset H \subset V'$, where V' is the dual space of V. We use the notation (v_1, v_2) to indicate the value of $v_1 \in V'$ at $v_2 \in V$; this coincides with the scalar product in H, whenever $v_1 \in H$. The norms in V and H will be denoted by $\|\cdot\|_V$ and $\|\cdot\|_H$, respectively.

We suppose that

(i) The injection of V into H is compact.

Consider a linear continuous operator $A\colon V \to V'$, such that

$$\text{(ii)} \quad \begin{cases} (Au, u) \geqq c\|u\|_V^2\,, & c > 0\,, \quad \text{for all } u \in V\,,\\ (Au, v) = (Av, u)\,, & \text{for all } u, v \in V\,. \end{cases}$$

Let $f\colon [0,\infty] \to H$ satisfy

(iii) $f \in AC_{\text{loc}}([0,\infty); H)$,

where AC_{loc} stands for the space of all H-valued, locally absolutely continuous functions on $[0,\infty)$.

Finally, let $\eta\colon [0,\infty) \to R$ satisfy either the conditions

$$\text{(iV)} \quad \begin{cases} \eta \in AC_{\text{loc}}([0,\infty); R);\ \eta(0) > 0\,,\\ \eta' \text{ is of bounded variation, locally on } [0,\infty), \end{cases}$$

or

$$\text{(V)} \quad \eta \in C[0,\infty) \cap C^2(0,\infty);\ \eta(0) > 0\,,$$
$$(-1)^k a^{(k)}(t) \geqq 0\,, \qquad k = 0, 1, 2\,, \qquad t > 0\,.$$

The following result is a linearized version of ([5], Theorem 2.1). (See also Remark 2.5 and the Note added in proof in [5]),

Theorem 1. *Assume that* (i)–(iii) *and* (iV) (*respectively* (V)) *hold. Then the initial-value problem* (8) *has a solution* u, *satisfying*

$$u \in AC_{\text{loc}}([0,\infty); H) \cap L^\infty_{\text{loc}}([0,\infty); V)\,, \tag{9}$$

$$\frac{du}{dt} \in L^\infty_{\text{loc}}([0,\infty); H) \cap AC_{\text{loc}}([0,\infty); V')\,, \tag{10}$$

$$\frac{d^2u}{dt^2} \in L^1_{\text{loc}}([0,\infty); V')\,. \tag{11}$$

Remark. It follows easily that conditions (V) imply the uniqueness of the solution (cf. [6], Lemma 1).

We first apply Theorem 1 to the case of problem (4), (5). Set

$$V = H_0^1(0,d)\,, \qquad H = L^2(0,d)\,, \qquad V' = H^{-1}(0,d)\,, \tag{12}$$

and define $A\colon V \to V'$ by

$$Au = -\frac{d^2u}{dx^2}\,, \qquad u \in V\,, \tag{13}$$

where the derivative is taken in the sense of distribution theory. It is immediate that conditions (i) and (ii) hold.

Now observe that the unknown φ in (4) may be interpreted as a function of t with values in H; specifically, define $u\colon [0,\infty) \to H$ by

$$u(t)\,(x) = \varphi(t,x)\,, \qquad x \in (0,d)\,. \tag{14}$$

On account of (12)–(14) we can rewrite (4), (5) under the form (8), where f is also viewed as a map from $[0,\infty)$ into H. As a direct consequence of Theorem 1, we have

Theorem 2. *Let conditions* (iii) *and* (iV) (*or* (V)) *be satisfied. Then the problem* (4), (5) *has a solution, in the sense of* (9)–(14).

In the case of equations (6), (7) we choose

$$V = H_0^1(R_1, R_2)\,, \qquad H = L^2(R_1, R_2)\,,$$

with the norms in V and H given by

$$\|u\|_V^2 = \int_{R_1}^{R_2} r\left(\frac{du}{dr}(r)\right)^2 dr; \qquad \|u\|_H^2 = \int_{R_1}^{R_2} ru^2(r)\,dr\,.$$

Obviously, (i) is fulfilled. A straightforward argument then shows that the operators (with distributional derivatives)

$$A_n u = -\frac{d^2u}{dr^2} - \frac{1}{r}\frac{du}{dr} + \frac{n}{r^2}\,, \qquad u \in V\,, \qquad n = 0, 1\,, \tag{15}$$

are linear and continuous from V to V' and satisfy (ii).

As above, we may think of φ_n and f as of functions from $[0,\infty)$ into H and replace (6), (7) by

$$\frac{du_n}{dt}(t) + \int_0^t \eta(t-s)\, A_n u_n(s)\,ds = f(t)\,, \qquad t \geqq 0\,,$$
$$u_n(0) = 0\,, \qquad n = 0, 1\,, \tag{16}$$

with A_n given by (15) and $u_n(t)\,(r) = \varphi_n(t,r)$. Applying again Theorem 1 we obtain

Theorem 3. *If conditions* (iii) *and* (iV) (*resp.* (V)) *hold, then Eq.* (16) *has a solution* u_n *satisfying* (9)–(11).

Remark. The existence of solutions to problem (6), (7) can also be derived by Galerkin's method, under stronger conditions on η (cf. [7]).

References

1 FETECĂU, C., On the uniqueness of some flows in simple fluids, Acta Mech. **25**, 125 (1976).
2 FETECĂU, C., Non-steady helical flows of heat conducting simple fluids with memory, Rev. Roum. Sci. Techn.-Méc. Appl. **24**, 153 (1979).
3 TRUESDELL, C.; NOLL, W., The Non-Linear Field Theories of Mechanics, Handbuch der Physik, Vol. **III/3**, Springer Berlin/Heidelberg/New York 1965.
4 BARBU, V., Nonlinear Semigroups and Differential Equations in Banach spaces, Noordhoff, Leyden, 1976.
5 AIZICOVICI, S., On a nonlinear integrodifferential equation, J. Math. Anal. Appl. **63**, 385 (1978).
6 BARBU, V., Nonlinear integro-differential equations in Hilbert space, Confer. Sem. di Mat. Univ. Bari **143** (1976).
7 CHAURE, A., Ecoulement de Couette en viscoélasticité linéaire finie, J. Mécanique **16**, 67 (1977).

Eingereicht am 5. 11. 1980

Anschrift: Dr. SERGIU AIZICOVICI, Seminarul Matematic, Universitatea „Al. I. Cuza", Iaşi 6600, Romania

ZAMM 61, 661–662 (1981)

L. Petković / M. Petković

The Representation of Complex Circular Functions Using Taylor Series

The representation of functions, defined on the set of complex circular intervals $K(C)$, is considered. Using Taylor series and inclusive monotonicity, the inclusive interval function $F: D \to E$ $(D, E \subset K(C))$ for an analytical function f is introduced. F is the disc extension of the complex set $\{f(z) \mid z \in Z, Z \in K(C)\}$. The proposed representation, which has a centered form, is applied for finding the inclusive functions of standard (subroutine) functions.

Let $Z = \{z; r\}$, $z \in C$, $r \geqq 0$, be a disc with center at c and radius r. Denote the set of all discs in complex plane with $K(C)$.

Let H_1 and H_2 be subsets of C and let f: $H_1 \to H_2$. Denote the family of circular subsets of H_1 by $G(H_1)$, i.e. $G(H_1) \subset K(C)$, and the family of closed subsets of H_2 by $S(H_2)$. We will call f^*: $G(H_1) \to S(H_2)$ the *closed set-valued mapping* and

$$f^*(Z) = \{f(z) \mid z \in Z\} \quad \text{for each} \quad Z \in G(H_1)$$

the *closed united extension of* f. Thus, $f^*(Z)$ is the union in H_2 of all sets containing a single element $f(z)$ for some z in Z.

It is not always possible to determine the set value $f^*(Z)$ (See [3]), but when possible it is not simple. Often, the set $\{f(z) \mid z \in Z\}$ is not a disc, which makes difficulties in practical computation. In order to avoid this obstacle, we will define an interval function F in complex circular arithmetics as an interval extension F: $G(H_1) \to G(H_2)$ $(G(H_2) \subset K(C))$ of the function f^* such that $F(Z) \supseteq f^*(Z)$ for each $Z \in G(H_1)$. The set $F(Z)$ is a disc (discs) which incloses the closed region (regions) $\{f(z) \mid z \in Z\}$. Such an interval extension F will be called an *inclusive interval function.*

In this paper we will consider a class of inclusive interval functions which, from the general point of view, have a centered form (See, e.g. [1]). Let the argument of the function F be a disc $Z = \{c; r\}$ with the contour Γ. The closed region (or the regions when f is a multiform function, e.g. $f(z) = z^{1/n}$) $\{f(z) \mid z \in Z\}$ can be inclosed by the disc (or the discs) $F(Z)$ with center at point $f(c)$ and radius $R^* = \max_{z \in \Gamma} |f(c) - f(z)|$. Thus, instead of the set $\{f(z) \mid z \in Z\}$, the disc

$$F(Z) = \{f(c); R^*\} \supseteq f^*(Z) = \{f(z) \mid z \in Z\} \tag{1}$$

can be considered.

In the general case the maximal distance $|f(c) - f(z)|$ $(z \in \Gamma)$ cannot be determined, or the required procedure is complicated so that, sometimes, we adopt a radius $R \geqq R^*$. In this paper we will present a simple procedure for finding the radius R and consequently, the inclusive interval function $F(Z) = \{f(c); R\}$. This procedure uses Taylor series of analytical function (See [5]). In some cases the obtained radius is equal to R^*.

Theorem: *Let f be an analytical function on D, where $D \subset K(C)$, and let f map a disc $Z = \{c; r\} \in D$ on a closed region. Then, for the closed united extension f^* of f and its inclusive extension F we have*

$$f^*(Z) \subseteq F(Z) = \left\{f(c); \sum_{k=1}^{+\infty} \frac{|f^{(k)}(c)|\, r^k}{k!}\right\}. \tag{2}$$

Proof: Using a Taylor series of the analytical function f in the center $z = c$

$$f(z) = f(c) + \sum_{k=1}^{+\infty} \frac{f^{(k)}(c)}{k!} (z - c)^k, \qquad z \in Z, \tag{2}$$

and inclusive monotonocity, we obtain

$$f^*(Z) \subseteq F(Z) = f(c) + \sum_{k=1}^{+\infty} \frac{f^{(k)}(c)}{k!} (Z - c)^k$$

$$= f(c) + \sum_{k=1}^{+\infty} \frac{f^{(k)}(c)}{k!} \{0; r\}^k$$

$$= f(c) + \sum_{k=1}^{+\infty} \frac{f^{(k)}(c)}{k!} \{0; r^k\}$$

$$= \left\{f(c); \sum_{k=1}^{+\infty} \frac{|f^{(k)}(c)|\, r^k}{k!}\right\}.$$

Corollary: *Let $|f^{(k)}(c)| \leqq Q$, $k = 1, 2, \ldots$; Q is a real constant Then*

$$f^*(Z) \subseteq F(Z) \subseteq \{f(c); Q(e^r - 1)\}. \tag{3}$$

Proof: On the basis of (2) we have

$$f^*(Z) \subseteq F(Z) \subseteq \left\{f(c);\ Q \sum_{k=1}^{+\infty} \frac{r^k}{k!}\right\} = \{f(c); Q(e^r - 1)\}.$$

We will apply the formulae (2) and (3) for finding inclusive interval functions for standard (subroutine library) functions In realisation of the routines for computation of these functions (on FORTRAN IV) the subroutine functions ABS, SQRT, EXP, SIN, COS, ALOG and ATAN can be used.

1. $f(z) = z^n$.
According to (2) we have

$$Z^n = \{z^n \mid z \in Z\} \subseteq \left\{c^n; \sum_{k=1}^{n} \frac{n(n-1)\ldots(n-k+1)\,|c|^{n-k} r^k}{k!}\right\}$$

$$= \left\{c^n; \sum_{k=1}^{n} \binom{n}{k} |c|^{n-k} r^k\right\} = \{c^n; (|c| + r)^n - |c|^n\}.$$

The same result can be obtained applying multiplication of discs $Z_1 = \{c_1; r_1\}$ and $Z_2 = \{c_2; r_2\}$:

$$Z_1 \cdot Z_2 = \{c_1 c_2; |c_1|\, r_2 + |c_2|\, r_1 + r_1 r_2\},$$

which has been introduced by I. Gargantini and P. Henrici [2].

2. $f(z) = e^z$.

Since $\frac{d^k}{dz^k}(e^z) = e^z$, $k = 1, 2, \ldots$, using (3) we get

$$e^Z = \{e^z \mid z \in Z\} \subseteq \{e^c; |e^c|\,(e^r - 1)\}.$$

3. $f(z) = \sin z$.

Since

$$\left|\frac{d^{2k-1}}{dz^{2k-1}}(\sin z)\right| = |\cos z|,$$

$$\left|\frac{d^{2k}}{dz^{2k}}(\sin z)\right| = |\sin z|, \qquad k = 1, 2, \ldots,$$

using the formulae

$$|\cos c| = |\cos(x + iy)| = |\cos x \operatorname{ch} y - i \sin x \operatorname{sh} y| = (ch^2 y - \sin^2 x)^{1/2},$$

$$|\sin c| = |\sin(x + iy)| = |\sin x \operatorname{ch} y - i \cos x \operatorname{sh} y| = (\operatorname{ch}^2 y - \cos^2 x)^{1/2},$$

on the basis of (2) we obtain

$$\sin Z = \{\sin z \mid z \in Z\} \subseteq$$

$$\subseteq \left\{\sin c;\ |\cos c| \sum_{k=1}^{+\infty} \frac{r^{2k-1}}{(2k-1)!} + |\sin c| \sum_{k=1}^{+\infty} \frac{r^{2k}}{(2k)!}\right\}$$

$$= \{\sin c; \operatorname{sh} r\,(\operatorname{ch}^2 y - \sin^2 x)^{1/2} + (\operatorname{ch} r - 1)(\operatorname{ch}^2 y - \cos^2 x)^{1/2}\}.$$

4. $f(z) = \cos z$.
Using the consideration for the function $f(z) = \sin z$, we find

$$\cos Z = \{\cos z \mid z \in Z\} \subseteq$$

$$\subseteq \{\cos c; \operatorname{sh} r(\operatorname{ch}^2 y - \cos^2 x)^{1/2} + (\operatorname{ch} r - 1)(\operatorname{ch}^2 y - \sin^2 x)^{1/2}\}.$$

5. $f(z) = \ln z$.
Let $0 \notin Z = \{c; r\}$, i.e. $|c| > r$ holds. Since

$$\left|\frac{d^k}{dz^k}(\ln z)\right| = \frac{(k-1)!}{|z|^k}, \qquad k = 1, 2, \ldots,$$

and $\frac{r}{|c|} < 1$, applying (2) we obtain

$$\ln Z = \{\ln z \mid z \in Z, 0 \notin Z\} \subseteq$$

$$\subseteq \left\{\ln c; \sum_{k=1}^{+\infty} \frac{\left(\frac{r}{|c|}\right)^k}{k}\right\} = \left\{\ln c; -\ln\left(1 - \frac{r}{|c|}\right)\right\}.$$

6. $f(z) = \operatorname{arctg} z$.
Assume that $\pm i \notin Z = \{c; r\}$, i.e. the inequalities $|i - c| > r$ and $|i + c| > r$ are valid. We will use the formula

$$\operatorname{arctg} z = \frac{1}{2i} \ln \frac{i - z}{i + z}$$

and the inclusive interval function for $\ln Z$. We have

$$\operatorname{arctg} Z = \{\operatorname{arctg} z \mid z \in Z, \pm i \notin Z\} \subseteq$$

$$\subseteq \frac{1}{2i} \ln (i - Z) - \frac{1}{2i} \ln (i + Z) \subseteq$$

$$\subseteq \frac{1}{2i} \left\{\ln (i - c); - \ln \left(1 - \frac{r}{|i - c|}\right)\right\} -$$

$$- \frac{1}{2i} \left\{\ln (i + c); - \ln \left(1 - \frac{r}{|i + c|}\right)\right\}$$

$$= \left\{\frac{1}{2i} \ln \frac{i - c}{i + c}; - \frac{1}{2} \ln \left(1 - \frac{r}{|i - c|}\right) \left(1 - \frac{r}{|i + c|}\right)\right\}$$

$$= \left\{\operatorname{arctg} c; - \frac{1}{2} \ln \left(1 - \frac{r}{|i - c|}\right) \left(1 - \frac{r}{|i + c|}\right)\right\}.$$

The radius $R = - \frac{1}{2} \ln \left(1 - \frac{r}{|i - c|}\right) \left(1 - \frac{r}{|i + c|}\right)$ is larger than a radius which can be obtained by optimization of the distance $|f(c) - f(z)|$, $z \in Z$, (See [1]), but R is given by a much simpler expression and requires less numerical operations.

7. $f(z) = z^{1/n}$.
Assume that $0 \notin Z = \{c; r\} = \{|c| e^{i\theta}; r\}$, i.e. $|c| > r$ holds. The derivative of k-th order of the function $f(z) = z^{1/n}$ is given by

$$\frac{d^k}{dz^k} (z^{1/n}) = \frac{1}{n} \left(\frac{1}{n} - 1\right) \cdots \left(\frac{1}{n} - k + 1\right) z^{(1/n) - k}, \quad k = 1, 2, \ldots$$

Applying (2) we obtain

$$Z^{1/n} = \{z^{1/n} \mid z \in Z, 0 \notin Z\} \subseteq$$

$$\subseteq \left\{c^{1/n}; \sum_{k=1}^{+\infty} \frac{(-1)^{k-1} \frac{1}{n} \left(\frac{1}{n} - 1\right) \cdots \left(\frac{1}{n} - k + 1\right) |c|^{(1/n) - k} r^k}{k!}\right\}$$

$$= \left\{c^{1/n}; - \sum_{k=1}^{+\infty} (-1)^k \binom{\frac{1}{n}}{k} |c|^{(1/n) - k} r^k\right\} =$$

$$= \{c^{1/n}; |c|^{1/n} - (|c| - r)^{1/n}\}$$

$$= \bigcup_{m=0}^{n-1} \left\{|c|^{1/n} \exp \left(i \frac{\theta + 2m\pi}{n}\right); |c|^{1/n} - (|c| - r)^{1/n}\right\}.$$

The binomial series, which appears in this procedure, is convergent because of $\frac{r}{|c|} < 1$.

Note that the derived formula for the k-th root of a disc is also obtained in [4], but using (1). Thus, in the case of the inclusive interval function for $f^*(Z) = Z^{1/n}$, we have $R = R^*$.

According to the examples of the presented functions, it can be concluded that formula (2) enables simpler evaluation of inclusive interval functions compared to the known methods. This advantage is very remarkable in the case of more complex functions. In such cases the formula (3) can be applied effectively. The lower bounds of radii, resulting from (2), is not weaker than those obtained by some other methods (see, e.g. [1]). The values of these radii are good approximations of R^* and give satisfactory results in numerical calculation.

References

1 Börsken, N. C., Komplexe Kreis-Standardfunktionen, Freiburger Intervall-Berichte 78/2, Institut für Angew. Math. Universität Freiburg i. Br. 1978.
2 Gargantini, I.; Henrici, P., Circular arithmetic and the determination of polynomial zeros, Numer. Math. 18, 305–320 (1972).
3 Moore, R. E., Methods and Applications of Interval Analysis, SIAM Studies in Applied Mathematics, Philadelphia 1979.
4 Petković, M.; Petković, L J., On a representation of the k-th root in complex circular interval arithmetic, Interval Mathematics 1980 (ed. K. Nickel), Academic Press Proc., New York 1980, 473–479.
5 Reichmann, K., Die Konvergenz von Intervall-(Potenz-) Reihen mit Anwendungen auf Intervall-Anfangswertprobleme (Dr.-Dissertation), Freiburger Intervall-Berichte 80/4, Institut für Angew. Math. Universität Freiburg i. Br., 1980.

Eingereicht: 5. 1. 1981, korrigierte Fassung: 13. 4. 1981

Anschriften: Ljiljana D. Petković, Faculty of Mechanical Engineering; Dr. Miodrag S. Petković, Faculty of Electronic Engineering, University of Niš, Beogradska 14, 18000 Niš, Yugoslavia

ZAMM 61, 662–663 (1981)

G. Zielke

Verallgemeinerungen einer Testmatrix von Rutishauser mit exakten MOORE-PENROSE-Inversen

In der bekannten Arbeit „Once again: the least square problem" verwendete Rutishauser 1968 [2] zum Testen seines Verfahrens der doppelt relaxierten Pseudoinversen die (7×3)-Matrix

$$F_7 = \begin{bmatrix} 1 & 2 & 3 \\ 2 & 3 & 4 \\ 3 & 4 & 5 \\ 4 & 5 & 6 \\ 5 & 6 & 7 \\ 6 & 7 & 8 \\ 7 & 8 & 9 \end{bmatrix}$$

mit den Elementen $(F_7)_{i,k} = i + k - 1$, $i = 1(1)\,7$, $k = 1(1)\,3$. F_7 hat offensichtlich den Rang 2. Die Elemente der Moore-Penrose-Inversen F_7^+ gab Rutishauser exakt an:

$$(F_7^+)_{k,i} = \frac{1}{168} (-152 + 32i + 72k - 15ik).$$

Danach hat F_7^+ folgende Zahlenwerte

$$F_7^+ = \frac{1}{168} \begin{pmatrix} -63 & -46 & -29 & -12 & 5 & 22 & 39 \\ -6 & -4 & -2 & 0 & 2 & 4 & 6 \\ 51 & 38 & 25 & 12 & -1 & -14 & -27 \end{pmatrix}.$$

Die Matrix F_7 wurde auch von Molinari [1], einem ehemaligen Mitarbeiter von Rutishauser, zum Testen verschiedener Verfahren benutzt. In beiden Arbeiten [1] und [2] lieferten die besten Verfahren recht genaue Ergebnisse. Dies wird verständlich, wenn man die gute Kondition von F_7 berücksichtigt. Die singulären Werte von F_7 sind, auf 8 Stellen genau, $\sigma_1(F_7) = 24.922010$ und $\sigma_2(F_7) = 1.3760068$, und folglich ist $\operatorname{cond}_2(F_7) = ||F_7||_2 \cdot ||F_7^+||_2 = \sigma_1/\sigma_2 \approx 18.1$. Aus dieser kleinen Konditionszahl ist ersichtlich, daß F_7 eine sehr milde Testmatrix ist, die die Verfahren nicht strapaziert.

Um die Konditionszahl zu erhöhen, soll die Matrix F_7 nun auf zweierlei Weise verallgemeinert werden. Zunächst liegt es nahe, die Matrix auf n Zeilen zu vergrößern. Damit erhält man die $(n \times 3)$-Matrix

$$F_n = \begin{pmatrix} 1 & 2 & 3 \\ 2 & 3 & 4 \\ 3 & 4 & 5 \\ \vdots & \vdots & \vdots \\ n & n+1 & n+2 \end{pmatrix}$$

mit den Elementen $(F_n)_{i,k} = i + k - 1$, $i = 1(1)\,n$, $k = 1(1)\,3$. F_n hat natürlich ebenfalls den Rang 2, denn die 3. Spalte von F_n erhält man, wenn man vom Doppelten der 2. Spalte die 1. Spalte subtrahiert. Es läßt sich nun nachweisen, daß die Elemente von F_n^+ für alle natürlichen Zahlen $n > 1$ folgende Form haben

$$(F_n^+)_{k,i} = \frac{1}{n(n^2 - 1)} [-(4n^2 + 14n + 10) + (6n + 22)\, i +$$

$$+ (2n^2 + 6n + 4)\, k - (3n + 9)\, ik].$$

Die Konditionszahl steigt für wachsende n nur etwa linear an, und zwar mit einem Faktor zwischen 2.6 (bei $n = 7$) und 1.4. Für sehr große n gilt $\operatorname{cond}_2 (F_n) \approx \sqrt{2} \cdot n \approx 1.414\, n$. Einige Konditionszahlen sind: $\operatorname{cond}_2 (F_{50}) \approx 77$, $\operatorname{cond}_2 (F_{100}) \approx 148$,

$\text{cond}_2 (F_{500}) \approx 713$, $\text{cond}_2 (F_{1000}) \approx 1421$. Um eine Konditionszahl von etwa 10^5 zu erreichen, müßte $n > 70000$ gewählt werden. Es ist klar, daß Tests mit solchen Matrizen zu teuer sind.

Eine wirksamere Möglichkeit, die Konditionszahl zu erhöhen, besteht darin, die Matrixelemente mit einem oder mehreren frei wählbaren Parametern zu versehen, wobei aber nun die Ordnung der Matrix fest bleiben kann, vgl. [3]. Die RUTISHAUSERsche Testmatrix F_7 zum Beispiel kann durch Einführung eines additiven reellen oder komplexen Parameters a wie folgt verallgemeinert werden:

$$F_7(a) = F = \begin{pmatrix} a+1 & a+2 & a+3 \\ a+2 & a+3 & a+4 \\ a+3 & a+4 & a+5 \\ a+4 & a+5 & a+6 \\ a+5 & a+6 & a+7 \\ a+6 & a+7 & a+8 \\ a+7 & a+8 & a+9 \end{pmatrix},$$

$$(F)_{i,k} = a + i + k - 1\,, \qquad i = 1(1)\,7\,, \qquad k = 1(1)\,3\,,$$

$$\text{Rang}(F) = 2\,.$$

Nach der in [4] beschriebenen Methode erhält man als MOORE-PENROSE-Inverse

$$F^+ = \frac{1}{168} \times$$

$$\begin{pmatrix} -9a-63 & -6a-46 & -3a-29 & -12 & 3a+5 & 6a+22 & 9a+39 \\ -6 & -4 & -2 & 0 & 2 & 4 & 6 \\ 9a+51 & 6a+38 & 3a+25 & 12 & -3a-1 & -6a-14 & -9a-27 \end{pmatrix}.$$

Die singulären Werte $\sigma_1(F)$ und $\sigma_2(F)$ hängen natürlich vom Parameter a ab, $\sigma_3(F) = 0$. Für $|a| \gg 1$ ergibt sich $\sigma_1 \approx \sqrt{21}\,|a| \approx 4.5825757\,|a|$, $\sigma_2 \approx \sqrt{56}/|a| \approx 7.4833148/|a|$. Die Konditionszahl, die nunmehr im wesentlichen quadratisch vom Parameter a abhängt, beträgt damit für große $|a|$:

$$\text{cond}_2 (F) = \sigma_1/\sigma_2 \approx \sqrt{3/8}\,|a|^2 \approx 0.612\,|a|^2\,.$$

Für einige Werte von a ergeben sich folgende singuläre Werte und Konditionszahlen

a	σ_1	σ_2	$\text{cond}_2(F)$
0	$2.4922010\ 10^1$	$1.3760068\ 10^0$	$1.81\ 10^1$
1	$2.9199670\ 10^1$	$1.1744262\ 10^0$	$2.49\ 10^1$
10	$6.9446067\ 10^1$	$4.9380560\ 10^{-1}$	$1.41\ 10^2$
100	$4.8127227\ 10^2$	$7.1254587\ 10^{-2}$	$6.75\ 10^3$
500	$2.3142219\ 10^3$	$1.4818310\ 10^{-2}$	$1.56\ 10^5$
1000	$4.6054992\ 10^3$	$7.4460672\ 10^{-3}$	$6.19\ 10^5$
5000	$2.2935793\ 10^4$	$1.4951676\ 10^{-3}$	$1.53\ 10^7$
10000	$4.5848671\ 10^4$	$7.4795748\ 10^{-4}$	$6.13\ 10^7$

Literatur

1 MOLINARI, L., Vermittelnde Ausgleichung nach der Methode der kleinsten Quadrate. Prozedur VERMAG. In: GANDER, W.; MOLINARI, L.; ŠVECOVÁ, H. (Hrsg.), Numerische Prozeduren. Aus Nachlaß und Lehre von Prof. Heinz Rutishauser. Birkhäuser Verlag, Basel-Stuttgart 1977, 39–76.

2 RUTISHAUSER, H., Once again: the least square problem, Linear Algebra Appl. **1**, 479–488 (1968).

3 ZIELKE, G., Testmatrizen mit freien Parametern, Computing **15**, 87–103 (1975).

4 ZIELKE, G., Beiträge zur Theorie und Berechnung von verallgemeinerten inversen Matrizen, Dissertation B., Martin-Luther-Universität Halle-Wittenberg, Fakultät für Naturwissenschaften des Wiss. Rates, Halle 1977, 325 S.

Eingereicht am 4. 4. 1981

Anschrift: Dozent Dr. sc. GERHARD ZIELKE, Martin-Luther-Universität Halle-Wittenberg, Sektion Mathematik, Weinbergweg 17 DDR-4020 Halle.

BUCHBESPRECHUNGEN

Meschkowski, H., Mehrsprachenwörterbuch mathematischer Begriffe, deutsch-englisch-französisch-russisch-italienisch, Mannheim-Wien-Zürich, B.I.-Wissenschaftsverlag 1972, 135 S., ISBN 3-411-01409-1

Das vorliegende fünfsprachige (mathematische) Wörterbuch ist als Ergänzung zum „Mathematischen Begriffswörterbuch" desselben Autors in derselben Reihe gedacht und bringt — nach Möglichkeit, d. h. in nahezu allen Fällen — die adäquaten Übersetzungen für etwa 1500 Begriffe in den im Titel angegebenen fünf Sprachen. Es ist auf den deutschsprachigen Leser zugeschnitten, d. h. es besteht aus einem Teil, in dem die deutschen Begriffe in die anderen Sprachen übertragen werden, und einem zweiten, in dem die Begriffe aus den vier Fremdsprachen auf die jeweiligen deutschen zurückgeführt werden. Irgendwelche grammatische Erklärungen kann man nicht erwarten und findet man auch nicht. Wieweit Adäquatheit bei den russisch-sprachigen Begriffen erreicht wurde, ist bei derartigen Büchern oft interessant. Nach Stichproben zu urteilen, gibt es hier kaum Probleme, wenn auch z. B. für ‚Polyeder' in der Literatur ‚многогранник' bevorzugt wird. — Was das ‚mathematische Spektrum' betrifft, so wird man sich als Rezensent für die ZAMM natürlich nach Stichworten aus der angewandten Mathematik umsehen. Hier wird der Leser doch enttäuscht sein, denn es fehlen viele Begriffe, wie etwa ‚Methode der finiten Elemente' und damit zusammenhängende Begriffe, ‚Pivotelement', um einige herauszugreifen. Besonders auffällig war dem Referenten dieses Fehlen bei Begriffen aus der Mathematischen Programmierung. Hier vermißt man selbst elementare Begriffe wie ‚Zielfunktion', ‚Nebenbedingung', 'ober-, und ‚unterhalbstetig', ‚Schnittebene' usw. usw. — Der Referent möchte dies dem Autor nun nicht zum Vorwurf machen, da man bei den 1500 Begriffen einfach zu einer Auswahl gezwungen ist. Das ist ja auch geschehen: klassische und moderne Gebiete der reinen Mathematik, wie etwa Algebra, Logik, Zahlentheorie, Elementargeometrie, Topologie und Differentialgeometrie sind recht gut repräsentiert, und so ist dieses Buch wohl auch in erster Linie Lesern bzw. Nutzern zu empfehlen, die auf solchen Gebieten arbeiten.

Berlin H. WEINERT

Martini, R. (ed.), Geometrical Approaches to Differential Equations, Proceedings, Scheveningen, The Netherlands 1979, Lecture Notes in Mathematics 810, Berlin-Heidelberg-New York, Springer Verlag 1980, VII, 339 S., 64 Abb., DM 34,50. US $ 20.40. ISBN 3-540-10018-0

Den begrifflichen Schwerpunkt dieses Bandes — insbesondere der acht Hauptvorträge — stellt die Bäcklundtransformation (BT) dar. "There is no generally accepted definition of Bäcklund transformation". So schrieb MIURA (1976) in der Einleitung eines der BT gewidmeten Konferenzbandes [1]. Der vorliegende Band zeugt davon, daß in der Zwischenzeit ein deutliches Maß an begrifflicher Klärung erreicht worden ist.

Der einleitende Artikel von ESTABROOK gibt eine gut lesbare Einführung in Sprache und Technik der modernen Differentialgeometrie und führt über Cartans Theorie partieller Differentialgleichungen zu Prolongationsstrukturen und BTn. Der anschließende Beitrag von CORONES arbeitet die Beziehungen zwischen Solitonen, Integrabilitätsbedingungen und Liegruppen heraus. DODD und MORRIS bieten eine umfassende Theorie der BTn in Jet-Bündel-Formulierung mit zahlreichen Beispielen (Massives Thirring-Modell, 0(3) σ-Modell, Ernstgleichung). Eine Gegenüberstellung mit verwandten Arbeiten von PIRANI u. a. [2] wäre vom Standpunkt des Lesers sehr zu begrüßen gewesen. In einem zweiten Beitrag bestimmen DODD und MORRIS verallgemeinerte BTn für das AKNS (= Ablowitz, Kaup, Newell, Segur)-System sowie für allgemeinere mehrdimensionale Streuprobleme. Ein Artikel von JEFFREYS untersucht die Wechselbeziehungen zwischen BT, Riccati-Gleichungen und inverser Streumethode und erörtert Näherungsverfahren, die von einem gegebenen Differentialgleichungssystem eventuell zu einem durch die inverse Streumethode behandelbaren führen. CHRISTIANSEN stellt BTn für die sinus-Gordon-Gleichung in $2+1$ und in $3+1$ Dimensionen vor sowie entsprechende N-Solitonen-Lösungen. Ein damit eng verwandter Artikel ist in unserer Zeitschrift erschienen [3].

Leider vermißt der Leser in den untereinander thematisch verwandten Arbeiten eine wechselseitige Bezugnahme.

Der übrige Teil des Bandes ist thematisch recht heterogen und behandelt u. a. Pfaffsche Systeme, Hamilton-Systeme, Bedingungen für Separation der Variablen, Grenzzyklen und periodische Lösungen selbstgravitierender Systeme.

Literatur

1 Miura, R. (Herausg), Bäcklund Transformations, Lecture Notes in Mathematics No. 515, Springer-Verlag, Berlin, New York 1976.

2 Pirani, F. A. E.; Robinson, D. C.; Shadwick, W. F., Math. Phys. Stud. *1* (1979) S. 1–132.

3 Christiansen, P. L., ZAMM *60* (1980) S. T 242–T 243.

Berlin H. STEUDEL

Gaier, D., Vorlesungen über Approximation im Komplexen, Basel-Boston-Stuttgart, Birkhäuser Verlag 1980, 173 S., sFr. 26.–, DM 28,–, ISBN 3-7643-1161-4

Dieses Buch ist sehr gut sowohl als Einführung in dieses Gebiet geeignet, als auch als Überblick über neuere Entwicklungen; dies aus folgenden Gründen. Das Buch hat nicht den Ehrgeiz (wie leider heutzutage vielfach eingerissen), den Leser gleich am Anfang durch die ganze Wucht der Allgemeinheit mutlos zu machen. Es erfolgt immer eine kurze Heranführung des Lesers an die Fragestellung, was durchaus nicht eine gewisse durchdachte Kürze stört. In der Stoffauswahl erfolgt eine maßvolle Beschränkung, wobei aber alle wesentlichen Entwicklungen (auch neueste) irgendwie Berücksichtigung finden.

Der Inhalt in Kürze: Darstellung komplexer Funktionen durch Orthogonalreihen (Bergmannsche Kernfunktion, Darstellung konformer Abbildungen, Entwicklung nach Faberpolynomen). Approximation durch Interpolation (Hermitesche Interpolation, Satz von RUNGE u. ä.). Allgemeine Approximation im Komplexen (Satz von MERGELYAN, Ergebnisse von A. ROTH). Approximation auf abgeschlossenen Mengen (Sätze mehr theoretischer Tendenz). Anwendungen (z. B. Konstruktion regulärer Funktionen mit gegebenem Randverhalten u. ä.). Fast 200 Literaturhinweise.

Den derben typographischen Schnitzer auf dem Umschlag hat dieses Buch wahrlich nicht verdient.

Halle R. KÜHNAU

Hinton, E. / Owen, D. R. J., Finite Element Programming, Computational Mathematics and Applications, London, Academic Press 1977, XII, 305 S., $ 18.00. ISBN 0-12-349352-8 p/b

Man muß den Autoren zustimmen, daß es von der Theorie der Finiten Elemente ein weiter Weg ist bis zur Erstellung leistungsfähiger Programme für Rechner. Mit ihrem 1977 in erster Auflage erschienenen Buch wollen sie den Leser befähigen, eigene Programme für seine Aufgaben zu schreiben.

Kenntnisse in FORTRAN-Programmierung und Numerischer Analysis, speziell der Methode der Finiten Elemente werden vorausgesetzt. Das Buch enthält eine detaillierte Beschreibung dreier Programme zur Berechnung von Balken, Scheiben und Plattenbiegung.

Einheitlich werden für parabolische isoparametrische Elemente die Berechnung der Elementsteifigkeitsmatrizen, ihre Zusammenfassung zur Gesamtsteifigkeitsmatrix und die Berechnung des Lastvektors demonstriert. Das entstehende Gleichungssystem wird über die Frontlösungsmethode aufgelöst. Besonderer Wert wird auf die Erklärung von Ein- und Ausgabefunktionen sowie naheliegende Fehlerreaktionen gelegt.

Der Leser kann sich mit der Umsetzung eines numerischen Verfahrens in ein Programm vertraut machen. Der Wert des Buches ist in der Ausführlichkeit der Erklärung aller Teilschritte zu suchen, so daß es manchem eine willkommene Erfahrung zum Schreiben von Programmen für Finite-Elemente-Verfahren vermitteln kann.

Berlin G. LIPPOLD

Giloi, W. / Liebig, H., Logischer Entwurf digitaler Systeme. 2., überarb. Aufl., Hochschultext, Berlin-Heidelberg-New York, Springer-Verlag 1980. XII, 328 S., DM 48,–. US $ 28.40. ISBN 3-540-10091-1

Die erste Auflage erschien 1973. Seitdem hat die technische Entwicklung der Digitaltechnik wesentliche Fortschritte gemacht, die sich u. a. in den hochintegrierten Bausteinen der Speicher (ROM, RAM), PLA, Mikroprozessoren, SIO und PIO darstellen. Diese Entwicklung bestätigt die Konzeption des Werkes, auf einer etwas allgemeineren Basis die Grundprinzipien zu behandeln. Dabei wird eine relativ einheitliche mathematische Beschreibung für kombinatorische und sequentielle Schaltungen benutzt. Dem wird das stärker strukturelle Konzept der technischen Schaltungen zugeordnet. Diese deutliche Trennung vermeidet auch weitgehend Probleme, die sonst durch die nicht klare Trennung von Funktion und Struktur bei Schaltungen entstehen. Somit liegt ein Lehrbuch vor, das relativ klar und gut didaktisch die verschiedenen digitalen Beschreibungsmittel darstellt.

In den sechs Kapiteln werden behandelt: Grundlagen der Booleschen Algebra, Boolesche Algorithmen (im wesentlichen sequentielle Schaltungen, Automatentheorie), dann deren technische Realisierungen (Schaltnetze, Schaltketten, asynchrone und synchrone Schaltwerke). In der neuen Auflage wurde hierbei auch der Mikroprozessor einbezogen. Hierdurch mußte der Anhang zu Grundlagen der Automatentheorie entfallen. Mit diesen Änderungen konnte das Werk auf den neuesten Stand der Technik gebracht werden und wird so weiterhin als gut durchdachtes Einführungswerk vielen Lesern als wertvolles Handwerkszeug nützlich sein.

Berlin H. VÖLZ

Osaki, S. / Nishio, T., Reliability Evaluation of Some Fault-Tolerant Computer Architectures. Lecture Notes in Computer Science 97. Berlin-Heidelberg-New York, Springer-Verlag 1980. VI, 129 S., 46 Abb., DM 18,–. US $ 10.70. ISBN 3-540-10274-4

Im ersten Kapitel werden die verschiedenen Begriffe und mathematischen Definitionen der Zuverlässigkeit eingeführt: Zuverlässigkeit als Funktion der Betriebszeit, als mittlere Zeit bis zum ersten Fehler (MTFF), zwischen zwei Fehlern (MTBF), Ausfallzeit (MDT) und als Verfügbarkeit. Dabei werden verschiedene Arten der Redundanz, reparierbare und nicht reparierbare Systeme, sowie Übertragung dieser allgemeingültigen Begriffe auf die Besonderheiten der Rechentechnik betrachtet. Diese Einführung in die Zuverlässigkeit ist so angelegt, daß sie allgemein interessieren dürfte. In vier weiteren Kapiteln erfolgt dann die spezifische Anwendung auf Duplex-, Dual- und Multisysteme und zwar zunächst im normalen Betrieb und dann mit automatischer Fehlerunterdrückung (coverage-related). Weiter werden behandelt: Untersuchungen zum Informationsdurchsatz und zur Fehlerhaftigkeit von Ergebnissen, sowie ein hybrides System, das aus zwei Multisystemen und einem Reservesystem besteht.

Es werden jeweils die Grundstrukturen eingeführt, das Modell aufgestellt und dann die Zuverlässigkeitswerte berechnet. Die numerischen Beispiele werden in Kurven des prinzipiellen

Zeitverlaufs anschaulich dargestellt und so miteinander verglichen. In einem Anhang werden Markoffsche Erneuerungsprozesse behandelt.

Berlin H. Völz

Jones, N. D. (ed.), Semantics-Directed Compiler Generation, Proceedings of a Workshop, Aarhus, Denmark, 1980; Lecture Notes in Computer Science 94, Berlin-Heidelberg-New York, Springer-Verlag 1980. V, 489 S., 34 Abb., DM 45,50. US $ 26.90. ISBN 3-540-10250-7

Die Compilertechnik war von jeher eine Teildisziplin, in der theoretische Untersuchungen und praktische Anwendungen eng verbunden sind. Die fortschreitenden theoretischen Erkenntnisse stimulieren vor allem einen höheren Automatisierungsgrad in der Compiler-Entwicklung. So entstand eine beträchtliche Anzahl sogenannter „Compiler-Generatoren", die auf der Grundlage der formalen Definition der Programmiersprachen einen entsprechenden Compiler oder wesentliche Teile desselben automatisch erzeugen. Heute kann die automatische Erzeugung der Syntaxanalyse (für kontextfreie Sprachen) als praktisch befriedigend gelöst betrachtet werden. Probleme gibt es noch mit der Einbeziehung von Kontextbedingungen und der Code-Erzeugung, also Fragestellungen, die mit der Definition der Semantik gekoppelt sind. In den letzten 10 Jahren gab es nun auch hierbei entscheidende theoretische Impulse, vor allem durch algebraische Methoden. Im vorliegenden Sammelband wird darauf eingegangen, wie die verschiedenen theoretischen Zugänge zur Definition der Semantik für die Compiler-Entwicklung genutzt werden können.

Der Band ist in vier Teile gegliedert: 1. *Compilers based on denotational semantics*: Drei Beiträge geben Hinweise dazu, wie aus der „denotational" Semantik Regeln zu Erzeugung von Compilern abgeleitet werden können. 2. *Compiling and algebraic semantics*: In vier Beiträgen wird vor allem die Frage der Korrektheit eines Compilers und deren Nachweis auf der Grundlage abstrakter algebraischer Spezifikationen diskutiert. 3. *Attribute and affix grammars*: Attribut- (bzw. Affix-) Grammatiken sind eines der ältesten als praktisch relevant erkannten Instrumente zur Erfassung der Semantik. In 7 Beiträgen wird detaillierter auf die Nutzung von Attribut- und Affix-Grammatiken eingegangen. 4. *Related topics*: Von den drei Beiträgen dieses Teils nutzen zwei den automatentheoretischen Zugang (operationelle Semantik). Der letzte Beitrag deutet die formale Definition der Programmiersprache ADA mittels der Methode der „denotational" Semantik an.

Es zeigt sich, daß das Grundkonzept der Compilertechnik — Zuordnung von Bedeutung zu den syntaktischen Einheiten der Sprache — überall beibehalten wird. Vor allem durch höhere Abstraktion wurden neue Ergebnisse erzielt, die einen weiteren Schritt der automatischen Erzeugung von Compilern ermöglichen. Dabei ist es erfreulich, daß in den meisten Artikeln das Anliegen relativ ausführlich dargelegt wird und damit dem Leser ausreichende Informationen zur eigenen Nutzung der Methoden und Verfahren bereit stehen.

Dresden P. Bachmann

Marlin, C. D., Coroutines — A Programming Methodology, a Language Design and an Implementation, Lecture Notes in Computer Science 95, Berlin-Heidelberg-New York, Springer-Verlag 1980, XII, 246 S., 55 Abb., DM 29,50. US $ 17.40. ISBN 3-540-10256-6

This book presents a framework for programming with coroutines. The coroutine features of a variety of programming languages are compared using informal data control and sequence control models. This comparison yields the design goals for a PASCAL-based language called ACL (*A Coroutine Language*).

ACL's sequence control model can be regarded as a further elaboration of SIMULA's quasi parallel systems by introducing a clear distinction between generators, procedures and coroutines. The sequence control features support a programming methodology wich produces hierarchically structured coroutine programs, using the notions of caterval and elementary subprograms.

An ACL implementation is described giving hints on implementation techniques and storage management. Two appendices present the syntax diagrams and ACL programs for the solution of problems typical for coroutine programming.

The book is well written and contains an extensive bibliography on language design and coroutines. It deserves attention as an investigation in programming methodology, as an attempt to give a unified model of the semantics of coroutine features in different languages and as a semantics based approach to language design.

Berlin M. Hartwig

Bjørner, D. / Oest, O. N. (eds.), Towards a Formal Description of Ada. Lecture Notes in Computer Science 98. Berlin-Heidelberg-New York, Springer-Verlag 1980. XIV, 630 S., DM 46,—. US $ 27.20. ISBN 3-540-10283-3

Ada ist eine Programmiersprache, die den hohen Ansprüchen an eine moderne, universelle Programmiersprache gerecht wird. Sie enthält neben bekannten und bewährten Konzepten auch neue Mittel zur Steuerung paralleler Prozesse. Ada fußt auf der Sprache PASCAL und ihrer Nachfolger sowie auf Forschungen zu abstrakten Datentypen. Die Definition der Sprache ist sehr genau und gestattet sowohl die Ausnutzung von Maschineneigenschaften für Systemprogramme als auch die Produktion maschinenunabhängiger, portabler Software. Besonderer Wert wird darauf gelegt, daß Programme übersichtlich, gut strukturiert und somit wenig fehleranfällig geschrieben werden können und sich in effiziente Maschinenprogramme übersetzen lassen. Die Anforderungen an einen Compiler sind dementsprechend hoch. Die Implementatoren von Ada werden daher zur Entwicklung und Erprobung neuer Methoden der Software-Produktion herausgefordert. Das Danish Datamatics Centre (DDC) wendet zur Herstellung eines Ada-Compilers die im Wiener IBM-Laboratorium erarbeitete Entwicklungsmethode (VDM) an, deren Grundlage Denotational Semantics und Lambdakalkül bilden. Die Methode „basiert auf der Idee, daß eine Implementation, wenn nicht automatisch, so doch wenigstens systematisch, von einer formalen Definition der Quellsprache und einer Definition der Zielsprache ... abgeleitet werden kann". Diese Definitionen bilden den Hauptteil des Buches und stammen von J. Bungaard, O. Dommergaard, H. H. Løvengreen, J. Storkbank Pedersen und L. Schultz. Weiterhin findet man lesenswerte Ausführungen der Herausgeber über das Ada-Compiler-Projekt im DDC.

Berlin W. Issel

Bandemer, H. / Näther, W., Theorie und Anwendung der optimalen Versuchsplanung II, Handbuch zur Anwendung, Mathematische Monographien 48, Berlin, Akademie-Verlag 1980, XII, 590 S., 86,— M. BN 7623192 (6215/2)

Der vorliegende zweite Band schließt sich nahtlos an seinen Vorgänger an und beantwortet viele Fragen, die insbesondere für den Praktiker von großem Interesse sind. Die Darstellung ist auf einem mathematischen Niveau gehalten, die es Anwendern vieler Anwendungsgebiete gestattet, Nutzen aus den vorgeschlagenen Methoden zur Versuchsplanung zu ziehen, ohne daß grundlegende theoretische Erörterungen (wie im Band I enthalten) herangezogen werden müssen. Damit legen die Autoren ein didaktisch ausgereiftes, mit vielen Beispielen angereichertes Werk für die Praxis vor, das auch zahlreiche neuere Ergebnisse der Schätzung in linearen Regressionsmodellen berücksichtigt.

Berlin. H. Toutenburg

Prabhu, N. U., Stochastic Storage Processes, Queues, Insurance Risk, and Dams, Applications of Mathematics 15, Berlin-Heidelberg-New York, Springer-Verlag 1980, XII, 140 S., DM 38,—. US $ 22.50. ISBN 3-540-90522-7

Der Autor dieses Buches gehört zu denjenigen Bedienungstheoretikern, die weiterhin die klassischen Grundmodelle unter den traditionellen Unabhängigkeitsannahmen untersuchen und mit verfeinerten mathematischen Methoden neue Resultate erzielen. Wie Prabhu im Vorwort schreibt, verschob sich für ihn seit der Veröffentlichung des bekannten Buches „Queues and

Inventories“ (1965) das Interesse von der Untersuchung der stochastischen Modelle zum Studium der zugehörigen Zufallsprozesse. In der vorliegenden Monographie verwendet er insbesondere die Fluktuationstheorie für zufällige Irrfahrten, Levi-Prozesse und Wiener-Hopf-Faktorisierungen. Damit gelangt er zu Aussagen über Maxima, Minima, Erstpassagen sowie zu Grenzwertsätzen. Behandelt werden das Wartemodell GI/G/1 sowie einfache Lagerhaltungsmodelle mit stetiger Zeit (u. a. Dammodelle) und eng damit zusammenhängende Versicherungsmodelle. — Das Buch ist sehr gut organisiert und aufgemacht, es enthält zu jedem Kapitel interessante Übungsaufgaben und bibliographisch-historische Kommentare und vermittelt auch dem Kenner neue Einsichten. Es ist keinesfalls ein Buch für tinwender, aber m. E. von großem Wert für Mathematiker, die Aefer in die Bedienungstheorie eindringen wollen, sowie für Spezialisten der Theorie stochastischer Prozesse, die sich für die Anwendung allgemeiner Theorien interessieren.

Freiberg D. Stoyan

Holzweißig, F. / Dresig, H., Lehrbuch der Maschinendynamik. Leipzig, VEB Fachbuchverlag 1979. 416 S., 240 Abb., M 26,—. BN 5464351

An den Technischen Universitäten bildet die Maschinendynamik in den Lehrplänen der konstruktiven Fachrichtungen einen wichtigen Bestandteil. Sowohl im Energiemaschinenbau (z. B. bei Kolben- und Turbomaschinen) wie im Verarbeitungsmaschinenbau (z. B. bei Werkzeugmaschinen, bei Druck-, Textil- und Verpackungsmaschinen), ferner bei Fördermaschinen, Landmaschinen und Fahrzeugen treten schwierige dynamische Probleme auf, durch deren Lösung meist wichtige Erkenntnisse für die jeweils zweckmäßigste Konstruktion gewonnen werden können. Das natürliche Rüstzeug derartiger Probleme bietet die Technische Mechanik. Darüber hinaus besteht heute ein dringendes Bedürfnis nach Aufstellung und Anwendung geeigneter programmierbarer Rechenverfahren. Auf beide Gesichtspunkte wird in der bisherigen Literatur meist zu wenig Bezug genommen. Das neue Lehrbuch schließt diese Lücken und liefert zur jeweiligen Fachliteratur wichtige Ergänzungen. Trotz des umfangreichen Stoffes bleibt das Buch praxisorientiert: Mit den konstruktiven Elementen und ihrer schwingungstechnischen Kennwertbestimmung wird der Leser eingehend vertraut gemacht; erst dann wird er an die theoretische Problematik herangeführt. Bei Behandlung der linearelastischen Schwingungsketten wäre auch eine kurze Darstellung des Baranow-Verfahrens, das sich durch große Anschaulichkeit auszeichnet, von Vorteil gewesen.

München H. Neuber

Nayfeh, A. H., Introduction to Perturbation Techniques. New York, John Wiley & Sons 1981. XIV, 519 S., £ 16.00. ISBN 0-471-08033-0

Eine genauere und wirklichkeitsnähere mathematische Modellierung von Vorgängen in Natur und Technik führt in der Regel auf nichtlineare Problemstellungen, so auf nichtlineare algebraische Gleichungen bzw. Differentialgleichungen, auf nichtlineare Randbedingungen u. ä. Derartige Aufgaben lassen sich aber nur schwer oder nicht mehr exakt lösen; so werden entweder lediglich die entsprechenden linearen Anteile beachtet oder Approximationstechniken benutzt. Da ersteres Vorgehen meistens nur grobe Näherungen liefert, ist in den letzten Jahrzehnten eine Reihe von analytischen und numerischen Verfahren zur näherungsweisen Lösung von nichtlinearen Aufgabenstellungen entwickelt, ausgebaut und erprobt worden. Als eine der wichtigsten Klassen analytischer Verfahren gehören dazu insbesondere die Störungsmethoden, die im vorliegenden Buch ausführlich behandelt werden.

Da dieses Buch als Einführung für Leser ohne besondere Vorkenntnisse konzipiert ist, werden alle benötigten Hilfsmittel zur Darstellung der Störungsmethoden eingeführt und an ausgewählten Beispielen erläutert. Dazu gehören u. a. trigonometrische Identitäten, Herleitung, Konvergenz und Operationen von asymptotischen Reihenentwicklungen, Ermittlung von Approximationslösungen für algebraische und transzendente Gleichungen, Berechnungsmöglichkeiten von Integralen (z. B. Reihenentwicklungen, partielle Integration, Laplacetransformation, Methoden der stationären Phase und des steilsten Abstieges), Lösungsmethoden und Lösbarkeitsbedingungen für lineare Gleichungssysteme und Systeme linearer gewöhnlicher Differentialgleichungen.

Die wesentlichsten analytischen Störungsmethoden — Potenzreihenentwicklung nach einem kleinen Parameter, Methode von Lindstedt-Poincaré, Methode der Renormalisierung, Methode des mehrfachen Maßstabes, Mittelungsmethode — werden ausführlich dargestellt sowie durchgehend zur Behandlung von zahlreichen nichtlinearen Problemstellungen wie Duffingsche Gleichung, linearer gedämpfter Schwinger, selbsterregte Schwinger, Systeme mit quadratischen und kubischen Nichtlinearitäten, allgemeine schwachnichtlineare Systeme, fremderregte Schwingungen der Duffingschen Gleichung, mehrfrequente Erregungen, Mathieusche Gleichung benutzt. Abgerundet werden die Ausführungen durch Abschnitte über lineare gewöhnliche Differentialgleichungen mit variablen Koeffizienten und gewöhnliche Differentialgleichungen mit einem großen Parameter.

Das Buch ist leicht verständlich und übersichtlich abgefaßt; alle Abschnitte enthalten zur Erläuterung des Vorgehens illustrative Beispiele und angemessene Übungen, so daß ein Einarbeiten für Mathematiker, Physiker, Ingenieure und Studenten dieser Richtungen, die derartige Methoden benutzen möchten, in einfacher Weise und in kurzer Zeit möglich ist.

Allerdings muß darauf hingewiesen werden, daß das Buch auf Forschungsergebnisse aus der englischsprachigen Literatur ausgerichtet ist und viele in der UdSSR und auch in der DDR zu dieser Problematik entstandene Arbeiten nicht berücksichtigt sind.

Berlin H. Friedrich

Bradbury, L. J. S. u. a. (eds.), Turbulent Shear Flows 2, Selected Papers from the Second International Symposium on Turbulent Shear Flows, Berlin-Heidelberg-New York, Springer-Verlag 1980, IX, 391 S., 310 Abb., DM 128,—. US $ 75.60. ISBN 3-540-10067-9

Die 27 Arbeiten, die für den vorliegenden Band ausgewählt worden sind, stellen etwa ein Viertel der Beiträge dar, die auf dem 2. Internationalen Symposium über turbulente Scherströmungen (2.—4. 7. 1979, Imperial College London) vorgestellt worden sind. Die, wie schon im Tagungsband des 1. Symposiums (1977), sehr klare und vor allem einheitliche Gestaltung der Aufsätze der verschiedenen Autoren geht sicher auf die Einflußnahme der Herausgeber zurück. Die Arbeiten sind thematisch in 5 Kapitel eingeteilt. Jedes Kapitel enthält eine Einleitung, die einen Überblick über die im Kapitel dargestellten Ergebnisse gibt und die Arbeiten auch mit anderen gegenwärtig aktuellen Untersuchungen in Verbindung bringt.

Im 1. Kapitel („Turbulenzmodelle“, 7 Beiträge, Einleitung: B. E. Launder) werden u. a. die Bestimmung von Wahrscheinlichkeitsdichten für Skalare in turbulenten Scherströmungen, gewisse Abschwächungen der Kolmogorovschen Hypothesen bei Lokalisotropie, Turbulenzmodelle mit zwei Zeitmaßstäben (Verallgemeinerung des (k, ε)-Modells) und die Simulation von dreidimensionalen Spots mittels gewisser Wirbelverteilungen untersucht. Es erscheint bemerkenswert, daß durch ein Turbulenzmodell mit zwei Zeitmaßstäben die Relation zwischen den Breitenzunahmen des ebenen und des runden Freistrahls wenigstens qualitativ richtig geliefert wird. Das übliche (k, ε)-Modell versagt dabei bekanntlich. Das 2. Kapitel („Wandnahe Strömungen“, 5 Beiträge, Einleitung: J. C. Rotta) bringt u. a. Arbeiten über den Einfluß der Stromlinienkrümmung auf turbulente Scherströmungen (Grenzschichten an parallel zur Achse angeströmten rotierenden drehsymmetrischen Körpern, Strömung längs einer konvexen Wand) sowie den Wärmeübergang in der Grenzschicht von einer linienförmigen Wärmequelle zu einer nahen festen Wand. Diese Arbeiten sind vorwiegend experimentell angelegt. Im 3. Kapitel („Komplizierte Strömungen“, 5 Beiträge, Einleitung: H. McDonald) geht es u. a. um Strömungen in Kanälen mit Wänden unterschiedlicher Rauhigkeit, dreidimensionale Grenzschichten an schräg angeströmten Rotationskörpern, die Umströmung eines zweidimensionalen Hindernisses am Boden eines ebenen Kanals und das Verhalten eines runden Strahls in einer von zwei parallelen ebenen Platten begrenzten Querströmung. Im 4. Kapitel („Kohärente Strukturen“, 6 Beiträge, Einleitung: P. Bradshaw) werden kohärente Strukturen in freien Scherschichten untersucht, z. B. in ebenen und axialsymmetrischen Mischungsschichten und in bestimmten Nachlaufströmungen. Die Arbeiten des 5. Kapitels

(„Geophysikalische Strömungen", 3 Beiträge, Einleitung: J. C. André) behandeln u. a. die Modellierung einer Gewitterbö mittels Mehrparametermodellen und die Ausbreitung von Beimengungen in bestimmten turbulenten Strömungen.

Der Tagungsband gibt einen sehr wertvollen Überblick über Ergebnisse auf dem Gebiet turbulenter Scherströmungen, die insbesondere in den einschlägigen Forschungszentren der USA, Japans, der BRD, Englands und Frankreichs in der letzten Zeit erzielt worden sind.

Berlin G. Seifert

Oswatitsch, K., Contributions to the Development of Gasdynamics. Selected Papers, Translated on the Occasion of K. Oswatitsch's 70th Birthday. Eds.: Schneider, W. / Platzer, M., Braunschweig, Vieweg 1980. XX, 429 S., 192 Abb., DM 136,–. ISBN 3-08452-9

Professor Klaus Oswatitsch belongs to the group of leading pioneers and teachers of fluid mechanics that actively participated in its rapid development in the last half century.

The idea to commemorate his 70-th birthday by making available, in English, a selection of his contributions not easily available is praiseworthy. The selection contains 24 papers under 5 headings:
1. Fundamentals and General Aspects of Fluid Mechanics; 2. Laws of Similarity and Equivalence; 3. Methods for Solving; Problems of Gasdynamics; 4. Applications of Gasdynamics; 5. Two-Phase Systems and Condensation.

The oldest paper of 1941, is on: „The Dispersion and Absorption of Sound in Clouds". The most recent one: „Possibilities and Limitations of the Theory of Linearization in Fluid Dynamics" is the 1969 memorial lecture dedicated to his teacher Ludwig Prandtl. A full list of 10 monographs and books as well as of 119 contributed papers is appended and the Scientific Career and Work of Klaus Oswatitsch is briefly but ably described by hisformer student prof. W. Schneider.

Despite of its high price this book deserves to be made available in libraries for students, engineers and scientists interested in fluid mechanics, particularly gasdynamics to whose development professor K. Oswatitsch contributed so much.

Warsaw W. Fiszdon

Meirovitch, L., Computational Methods in Structural Dynamics, Alphen aan den Rijn, Sijthoff & Noordhoff 1980, XIV, 439 S., ISBN 90-286-0580-0

Das vorliegende Buch gibt eine klar und anregend geschriebene Einführung in das Gebiet der Schwingungen mechanischer Systeme unter besonderer Berücksichtigung numerischer Lösungsverfahren. In wohlabgestimmter Auswahl wird der Bogen gespannt von mathematischen Grundlagen wie linearer Algebra oder numerischen Methoden zur Lösung von Eigenwertproblemen bis hin zu beispielhaften Anwendungen wie dem Flattern eines Flugzeugflügels oder der simultanen Iteration von Systemen mit vielen Freiheitsgraden. In der ersten Hälfte des Buches werden die Eigenlösungen und Resonanzen diskreter Schwingungssysteme behandelt. Die zweite Hälfte ist den Schwingungen kontinuierlicher Systeme und der Diskretisierung solcher Systeme gewidmet. Als Beispiele sind Stäbe, Saiten und Membrane behandelt. Einen breiten Raum nimmt die vergleichende Darstellung verschiedener Diskretisierungsmethoden ein, wie der Rayleigh-Ritzschen Methode, des Galerkinverfahrens, der Kollokationsmethode, der Methode der kleinsten Quadrate, der Momentenmethode und vor allem der Methode der finiten Elemente, die als Spezialfall der klassischen Rayleigh-Ritzschen Methode mit einigen Nachteilen und wesentlichen rechentechnischen Vorteilen charakterisiert wird. Eine Reihe durchgerechneter Beispiele zeigt die Benutzung linearer, quadratischer und kubischer Elemente. Den Abschluß bildet eine Einführung in die Behandlung großer Systeme und die Methoden ihrer Reduktion auf kleine Systeme.

Wie schon andere Bücher des verdienstvollen Autors, zeichnet sich dieses durch interessante Koppelung von einführenden, auf den Kern der verschiedenen Methoden hinzielenden Erläuterungen, repräsentativen numerischen Beispielen, die einen Vergleich verschiedener Methoden gestatten, und allgemeinen Einschätzungen der Brauchbarkeit aus. Im Sinne der Anregung "it is perhaps appropriate to distinguish between applied mathematics and 'applicable' mathematics, where the latter may be narrowly defined as the subject of applied mathematics that has survived the curse of high dimensionality" führt das Buch von der angewandten zur anwendbaren Mathematik hin.

Berlin G. Schmidt

Gladwell, G. M. L., Contact Problems in the Classical Theory of Elasticity. Mechanics of Elastic and Viscoelastic Solids 1. Alphen aan den Rijn, Sijthoff & Noordhoff Intern. Publ. 1980. 716 S., Dfl. 120.00. $ 60.00. ISBN 90-286-0440-5

Das vorliegende Lehrbuch eröffnet dem Leser einen großräumigen Überblick über das weite Feld der mit klassischen Methoden lösbaren Kontaktprobleme der linearen Elastizitätstheorie. Es werden vorwiegend ebene Probleme behandelt bis auf ein kurzes Schlußkapitel über dreidimensionale Aufgaben. Entsprechend der Zielstellung des Buches kommen vorwiegend Integraltransformations-Integralgleichungs- und funktionentheoretische Methoden zur Anwendung. Diese mathematischen Hilfsmittel werden in drei größeren Kapiteln ausführlich zusammengestellt, wobei der Autor geschickt Brücken zu den folgenden Anwendungen schlägt. Da das Lehrbuch einen enzyklopädieartigen Einblick in den obengenannten Problemkreis verschafft, lernt der Leser das Wichtigste über beinah alle Integraltransformationen der mathematischen Physik an Hand deren interessanten Anwendungen auf Kontaktprobleme der linearen Elastizitätstheorie kennen. Bei den vielfältigen Anwendungen der Theorie linearer Fredholmscher und singulärer Integralgleichungen geht der Verfasser auf interessante Approximationsverfahren ein und entwickelt die auf Muskhelischvili zurückgehende Theorie singulärer Integralgleichungen an Hand eines Randwertproblems für eine Halbebene mit endlich vielen Kontaktabschnitten.

Nach einer knappen Einführung im ersten Kapitel in die lineare Elastizitätstheorie werden im zweiten Kapitel elementare klassische Kontaktprobleme behandelt wie diejenigen von H. Hertz, Cerruti und Boussinesq. Die folgenden Kapitel beinhalten dann eine außerordentlich reichhaltige Übersicht über geschlossen lösbare ebene Kontaktprobleme zu Halbebenen, ebenen Ecken, Zylinderflächen, Platten und Balken unter den verschiedensten Bedingungen.

Der Wert des Buches liegt einerseits in der Reichhaltigkeit der Darstellung wie andererseits in der Gründlichkeit, mit der die Lösungstheorien dieser zum klassischen Bestand der linearen Elastizitätstheorie gehörigen Probleme bis in die heutige Zeit verfolgt werden. Davon zeugen die über 700 Literaturhinweise am Schluß. Das Buch ist ein wertvoller Führer in das Feld der klassischen linearen ebenen Elastizitätstheorie und setzt relativ wenig an mathematischen Vorkenntnissen voraus.

Leipzig H. Beckert

Bratteli, O. / Robinson, D. W., Operator Algebras and Quantum Statistical Mechanics. Vol. 1: C^* and W^*-algebras. Symmetry Groups. Decomposition of States. Texts and Monographs in Physics. Berlin-Heidelberg-New York, Springer-Verlag 1979. XII, 500 S., DM 68,–. US $ 37.40. ISBN 3-540-09187-4

Die vorliegende Monografie (erster Band einer auf zwei Bände angelegten Darstellung) ordnet sich vom mathematischen Standpunkt aus ein in die Reihe der Monografien über Operatorenalgebren (Neumark, Normierte Algebren, 2. Aufl. 1968; Dixmier, Algèbres de von Neumann, 2. Aufl. 1969; Dixmier, Les C^*-algèbres et leurs représentations, 1964; Sakai, C^*-Algebras and W^*-Algebras, 1971; Takesaki, Theory of Operator Algebras I, 1979). Die Theorie dieser Algebren wurde begründet durch die bekannten Arbeiten von Murray und v. Neumann, deren Ziel es ursprünglich war, Fortschritte auf dem Gebiet unendlichdimensionaler unitärer Gruppendarstellungen zu erzielen. Später wurden unter dem Einfluß der Theoretischen Physik (Relativistische Feldtheorie, Statistische Quantenmechanik) neue Schwerpunkte gesetzt, die zu Umstrukturierung, Neuorientierung, neuen Zielsetzungen, kurz, zu einem beträchtlichen Aufschwung der ganzen Theorie Anlaß gaben.

Die Verfasser haben sich nun in dieser Monografie das Ziel gesetzt, gerade diese neue unter dem Einfluß der Theoretischen

Physik in Gang gesetzte Entwicklung der Theorie der Operatoralgebren widerzuspiegeln, indem neuere Forschungsarbeit auf diesem Gebiet zusammen mit sog. Standardtheorie in einem einheitlichen Rahmen und in sich geschlossen dargestellt wird. Darüber hinaus soll die Darstellung einem über Basiswissen der Funktionalanalysis verfügenden Leser zugänglich sein, was die Darlegung ausführlicher Beweise einschließt. Eine solche Zielsetzung unterliegt natürlich gewissen Beschränkungen. Die Verfasser schränken die im Hintergrund stehende physikalische Motivation auf die Statistische Quantenmechanik ein (die relativistische Feldtheorie bleibt außer Betracht), d. h. die Theorie wird vornehmlich im Hinblick auf mögliche Anwendungen in diesem Gebiet aufgebaut.

Zweifellos haben die Verfasser ihr Ziel erreicht und auch die angestrebte Selbstkonsistenz der Darstellung weitgehend verwirklicht. Die Darstellung ist sehr gut lesbar (ob die Anlage von Beweisen in „Schachtelform", mit im Verlaufe eines Beweises gesondert formulierten und bewiesenen Lemmata, die Übersicht wirklich fördert, erscheint allerdings dem Referenten zweifelhaft).

Nach einer einführenden Strukturtheorie von C^*-Algebren (einschließlich Einführung von Darstellungen und Zuständen) folgen Begriffsbildungen und Struktursätze im Zusammenhang mit v. NEUMANN-Algebren (wobei diese von vornherein als Operatoralgebren über Hilberträumen definiert werden). Es schließen sich an eine Darstellung der Tomita-Takesaki-Theorie (Einführung von Modulargruppe und selbstdualen Kegeln), ein Abschnitt über quasilokale Algebren sowie verschiedene Resultate aus der Klassifikationstheorie von W^*- und C^*-Algebren (Chap 2). Der Abriß der Theorie einparametriger Operator(halb-) gruppen (Chap 3) fußt einerseits auf der Banachraum-Theorie und andererseits auf der sog. algebraischen Theorie, die eine ausführliche Untersuchung der sog. Derivationen einschließt (u. a. Charakterisierung „räumlicher Derivationen" (spatial derivations)). In Chap 4 wird zunächst allgemein die sog. barycentrische Zerlegung von Zuständen behandelt (Choquet-Theorie). Der wichtige Begriff des orthogonalen Maßes wird ausführlich erläutert. Speziell werden die Zerlegung eines Zustandes in reine Zustände (extremale Zerlegung) sowie die sog. zentrale Zerlegung (in Zustände, in denen die Invarianten spezifische Werte haben) und subzentrale Zerlegungen erörtert. Schließlich wird der Begriff des G-invarianten Zustands (G Symmetriegruppe) eingeführt und es wird die sog. „ergodische Zerlegung" eines Zustands in extremale G-invariante Zustände studiert (Existenz und Eindeutigkeit; Charakterisierung ergodischer Zustände; Zerlegung G-ergodischer Zustände in Zustände mit niedrigerer Symmetrie). Für die Herleitung der ergodischen Zerlegung spielt die sog. asymptotische Abelizität (asymptotic abelianess) eine entscheidende Rolle. Speziell behandelt werden lokalkompakte abelsche Gruppen. Zum Schluß wird die „zweite Linie" der Zerlegungstheorie (Zerlegung der Darstellungen der Algebren, sog. „räumliche Zerlegung" (spatial decomposition)), die den Anfangspunkt der Entwicklung der Theorie der Operatoralgebren mit markiert und bereits bei v. NEUMANN eine wesentliche Rolle spielte, dargestellt (unter Einschluß der Theorie der direkten Integrale) und mit der Zerlegungstheorie der Zustände zusammengeführt (was eine Charakterisierung der orthogonalen Maße einschließt).

Jedem Kapitel sind „Notes and Remarks" angeschlossen, die einen guten Einblick in die aktuelle Entwicklung des Gebietes vermitteln. Das Literaturverzeichnis ist reichhaltig und umfaßt ca. 300 Titel.

Dieses Buch kann Lesern mit guten Kenntnissen in Funktionalanalysis und Operatortheorie, die sich für moderne operatortheoretische Methoden interessieren, die aus Erfordernissen der Theoretischen Physik hervorgegangen sind und in der Theoretischen Physik ihre Anwendung finden, sehr empfohlen werden.

In einem nachfolgenden zweiten Band sollen Anwendungen der Theorie in der Statistischen Quantenmechanik dargestellt werden.

Berlin H. BAUMGÄRTEL

Garrido, L. (ed.), Systems Far from Equilibrium, Proceedings, Sitges 1980. Lecture Notes in Physics 132. Berlin-Heidelberg-New York, Springer-Verlag 1980. XV, 403 S., 105 Abb., DM 47,–. US $ 27.80. ISBN 3-540-10251-5

Faszinierende geordnete und kohärente Bewegungen werden in Nichtgleichgewichtssystemen verschiedenster Größenordnung in den letzten Jahren mit großer Vehemenz untersucht. Statistische Zugänge zur Beschreibung der Evolution von Fluktuationen in dissipativen Systemen spielen eine wesentliche Rolle. Phasenübergangsanalogien, Fokker-Planck- bzw. Langevin-Modelle zur Approximation solcher Nichtgleichgewichtssysteme (vor allem chemische Reaktionen, Laser, Transportprozesse, dynamische kritische Phänomene, Lichtstreuung, aber auch Probleme der Soziologie), Mastergleichungsansätze, renormierte Störungsschemata für Übergangsprozesse, deren Bifurkationsanalyse und feldtheoretische Erweiterungen von Kontinuumsmodellen dissipativer Prozesse sind einige Trends in der mathematischen Behandlung dieser der theoretischen Physik entlehnten Probleme, die in den Proceedings „Systems Far from Equilibrium" zur gleichnamigen Konferenz in Sitges (Spanien, 1980) enthalten sind. Ein großer Teil der Beiträge ist einem gegenwärtig sehr bedeutsamen Problem, dem Verständnis chaotischer Bewegung, am Beispiel des Einsetzens von Turbulenz gewidmet. Diese Beiträge vermitteln einen experimentellen Überblick und suchen einen statistischen Zugang für chaotische Bewegung bei raum-zeitlich wechselwirkenden Systemen.

Die Autoren sind bemüht, ihre Spezialprobleme verallgemeinernd einzuordnen, was einen Überblick über den aktuellen Stand in diesem jungen Forschungsgebiet, der von diesem Band repräsentiert wird, erleichtert.

Berlin E. MATTHÄUS

Haken, H. / Wolf, H. C., Atom- und Quantenphysik. Eine Einführung in die experimentellen und theoretischen Grundlagen. Berlin-Heidelberg-New York, Springer-Verlag 1980. XIV, 365 S., 228 Abb., DM 54,–. US $ 31.90. ISBN 3-540-09889-5

Als Ziel- und Ausgangspunkt des modernen physikalischen Denkens nimmt die Quantentheorie eine zentrale Stellung in den Studienplänen für angehende Physiker ein. Dieses schöne, reich mit Abbildungen versehene Buch kommt mit relativ geringem mathematischen Rüstzeug aus, ist damit weitgehend vom 2. Studienjahr ab lesbar. Angelehnt an den Gang der historischen Entwicklung, den Bogen spannend von den frühen Atomhypothesen bis zum Laser, bringt es in ihrer Wechselwirkung sowohl die wichtigsten experimentellen Ausgangspunkte der Quantenphysik als auch die darauf gegründeten theoretischen Begriffsbildungen, die so als notwendig und vernünftig erscheinen. Die ausführliche Darstellung kann auch Chemikern und Mathematikern empfohlen werden.

Rostock M. HEINRICH

Zeitfracht Medien GmbH
Ferdinand-Jühlke-Straße 7
99095 Erfurt, Deutschland
produktsicherheit@kolibri360.de